URBAN ORACLES

Other books in this series from Lumen Editions:

Too Late for Man
Essays by William Ospina

The Joy of Being Awake
by Hector Abad

Diary of a Humiliated Man
by Felix de Azua

Silk
Stories by Grace Dane Mazur

Dialogue for the Left and Right Hand
Poems by Steven Cramer

S
by Harry Matthews, Mark Polizzotti, and John Echenoz

Pursuit of a Woman on the Hinge of History
by Hans Koning

URBAN ORACLES

stories by
Mayra Santos-Febres

translated from the Spanish by
Nathan Budoff & Lydia Platon Lázaro

Lumen Editions
a division of Brookline Books

Copyright ©1997 by Brookline Books, Inc.

All rights reserved. No part of this work covered by the copyright hereon may be reproduced or used in any form or by any means—graphic, electronic, or mechanical, including photocopying, recording, taping, or information storage and retrieval systems—without the permission of the publisher.

ISBN 1-57129-034-6

Library of Congress Cataloging-in-Publication Information
Santos-Febres, Mayra, 1966–
   [Corporal correcto. English.]
   Urban oracles: stories by Mayra Santos-Febres ; translated from the Spanish by Nathan Budoff & Lydia Platon Lázaro.
     p. cm.
   ISBN 1-57129-034-6 (pbk.)
   I. Budoff, Nathan, 1962–   . II. Platon Lázaro, Lydia.
III. Title.
PQ7440.S26C613   1997
863--dc21   97-246
                                                           CIP

Printed in Canada by Best Book Manufacturers, Louiseville, Quebec.

10 9 8 7 6 5 4 3 2 1

Published by:
Lumen Editions
Brookline Books
P.O. Box 1047
Cambridge, Massachusetts 02238-1047

Contents

Broken Strand 1

Abnel, Sweet Nightmare 9

Night Stand 13

Resins for Aurelia 17

Urban Oracles 35

Marina's Fragrance 43

Brine Mirror 53

The Parks 59

Stained Glass Fish 65

Dilcia M. 71

Act of Faith 77

Oso Blanco 79

Mystic Rose 97

A Normal Day in the Life of Couto Seducción 101

The Writer 107

Broken Strand

A little girl and a father and a dream and a memory broken like a nose, at ten years old, with alcoholic breath on top of it all. There are days when you have to walk languid out on the street to forget, high-heeled out onto the street, with plastic nails and Bondo on your face like a freshly painted car. With sandpaper, you have to remake yourself at times, and the hot comb heating on the stove waiting for the split end, the broken strand, to repair it again, recompose it to a vivid red. The comb gradually takes on the neon color of blood.

You wait for the sunset as you sit in an invented house in Barrio Trastalleres. The cement rises like smoke; the cousins and brothers of Miss Kety's Beauty Parlor's owner have made it rise. Miss Kety knows how to comb hair and retouch straightenings from ten-year-olds on. Miss Kety has a broken nose from a stray fist, a map of burns on her forearm, and she will not use her hot iron to tame the nappy curls of little four-year-old black girls that carry rings in backyard weddings. —Oh, renaissance fairies, nymphets from the projects, who at four already know about alcoholic breath and dried-out condoms on the sidewalk, and want to straighten their hair so they won't be so black and so ugly and so low-class. At that age Miss Kety isn't willing to seat them on top of two telephone books, put Vaseline behind their ears, turn on the iron, and look for a bucket of water while she recounts the gossip and blows her broken nose and looks for a towel to wrap around their neck. "Yetsaida, now you have to sit still, baby, because the comb is hot and you could mess up the map I have on my forearm." That's why Yetsaida had to wait; she had to wait until she turned thirteen for a good straightening.

Sitting still, watching the sunset pass between the two tall, tall buildings of the Trastalleres project, Miss Kety's Beauty Parlor sits high in the sky like a cement bird that carries you toward beauty. It sits on a rise, on a curve that opens to the sun. You walk down from the bird straightened and beautiful. The wind plays with your hair just like in Breck dreams, like in Mirta de Perales' television ads.

"Oh, goddess of healthy hair, holy protector of follicles' shine and moisture, of the porous protein strands—keep me from sweating, because the dream ends there. If I sweat, the waves will flow again into their original curls."

Miss Kety has a broken nose, a map of the world on her left forearm, and soft rollers, "so my hair doesn't get ruined." She does her own hair; she straightens it with fine products she buys in the discount store at Stop 16 in Santurce. Once a week she uses a Tresemmé protein treatment, cuts the tips off and applies vitamins so her hair won't dry out like straw, like the manes of all the other women in Trastalleres. That's how she's different from all the others. Miss Kety, so elegant—she is black, yes. "But that beautiful head of straight hair, dyed Auburn Copper red, falling over her shoulders"—oh, and she keeps it long, that tall beauty, tall with the map of the world on her left forearm that testifies to her bravery and sacrifice. Her fame is legendary even though, like all the other women in the neighborhood, her nose has been broken by a fist.

You hope for your turn, hope for the sunset, hope that nobody notices the curls and the misguided nose, hope for the sweaty comb colored a vivid red, hope for the branding ritual that will return the dream that you are beautiful, that Nature has fixed everything in you, all of your good qualities, and sends Aeolus to play with your hair. You dream that Nature has told the guy who stole Jupiter's car to drive slowly past you so that even the sun can see your hair shining and straight as it should be. You touch your broken nose and know about the broken nose of Miss Kety, the savior, the emissary of true beauty.

Ten years old and a fed-up mother, a father arriving home and you waiting with curls and with homework from social studies class, a map of the world to color and turn in—Asia in yellow, Africa in purple, Central America, and Nicaragua, whose capital is Managua—a father who arrives and you show him the map and ask him for crayons and continue to ask, because the squeaky wheel gets the grease. The father doesn't even touch her curls out of disgust. Yes, disgust. Yetsaida has seen it in his face, in the jokes about his tangled hands every time he touches her hair, in Mama's irritation combing her hair with a broken-toothed comb—take that—with whose back she knocks her on the head—"Be quiet, girl! Sit still." She asks for crayons; she's going to turn in her map of the world and she's going to get an A so they take her to Miss Kety's, who's going to comb her hair, cut off the tips, take her to Managua in her flying bird beauty parlor and to Asia, pale yellow, just like she wants her hair to be. And straight, straight, straight. She insists to the father, who arrives from the street stumbling, who yells at her, to whom she yells back, who chases her to hit her and grabs her by the neck, who makes Mama cry and grab her nose just in case he throws a fist—"Whose nose?" All the screaming bewilders her so; the blood bewilders so, and those curls tight as snails, ah and the river that comes out of the nose, red, neon red like Miss Kety's comb.

The sun arrives, red, red, red; it overflows with blood like a ripening womb, like it has a deep gash in its system, like they have given it a tremendous beating. Now she is grown up; now she can wait her turn at Kety's Beauty Parlor and wear Press-on Nails and get three holes pierced in her ears. Now she shaves her legs and waits to make herself over just like she always knew that she should be. She looks at Miss Kety with the complicity of an initiate who had to learn to retouch her straightening alone, to make her own hair flip, to put on Hairweave, to mix the glue for plastic fingernails. She knows about tips and ammonias to decolor and she knows about multivitamin packets and that Porosity Control shampoos cause dandruff.

She discovered how Miss Kety learned to perfect herself. They are a secret society; she and Miss Kety have given their lives to the worship of true beauty, that which is real and appears on the screens and in the ads—which we know are there to make us buy Pampers and bad clothes from MeSalve, but they can all go to hell. Beauty is true. Somebody had to know to dye actresses' hair sun-red, such a good blood red. That's no trick of the camera, to make the red seem as if it were coming out of a deep gash, as if it were a hot comb shining in the most intense follicle. She wants to dedicate her life to this, beyond a beauty parlor in Trastalleres; she is going to enroll in the Beauty Academy next to the D'Rose Modeling School next year, oh yes, and in the Sky Academy of Looks and Beauty in Miami when she continues her studies. But first, yes, first, she has to let herself be made up by the ancient Miss Kety, her guardian of brilliance. She has to let herself be touched by Miss Kety's hands to really know what beauty is all about, to smell the map of the world on her left forearm. This is the advice of her inner voice, so she waits for the sunset.

"Oh, these roots, girl. Your hair's gonna break right off. Look at all these broken strands."

And she blows her broken nose, which the neophyte also blows; well she knows it was destined to be this way. The sun continues pouring down the walls of Trastalleres, and an unkempt girl rides her bicycle down the street. Miss Kety's hair is in a pentecostal bun dyed with Clairol Gel Colors Purple Blue Black without peroxide or ammonia. She is the perfect image of one who *knows* about rebellious curls, knows how to tame her own, and that's just how Yetsaida will look when she leaves the beauty parlor. She opens a bottle of Easystyle Products Hair Relaxer Without Lye, a little package of proteins. Miss Kety has put the hot comb away permanently; the map of the world on her forearm is now complete. Oh yes, and Yetsaida, the neophyte, accepts what comes next, the expert and unassertive hand which agitates her hair. She abandons herself to the pulling sensation, to the delicious feeling of fingers that don't want to break anything of hers, that want to leave her hair beautiful and radiant for the west wind to play with and make her a woman.

With an expert hand she runs the comb smeared with cream through Yetsaida's hair, which she deftly divides into four perfect squares. Miss Kety expertly makes her fingers dance strand by strand, threading a dream for the nappy curls that bleed in their own way. "This straightener doesn't hurt, it doesn't burn like the old ones, and it even conditions the hair." Progress and the future have brought forth the absence of fire, over which Yetsaida would have liked to pass like those crazy New Age gringos on the program *Ocurrió Así*, the ones who walked on burning coals. By crossing burning irons she wants to arrive at that peace—expert, gentle, straight, and without peroxide. With the sensation and the odor of something dying, a strand that screams, changing its texture, floating in the breeze without fear of the wet night air. Miss Kety's expert hand poking strand by strand, smearing cream into the hair so that no fugitive strands remain tight or curled (as sometimes happens with rookie beauticians), and all the while she's thinking about the absence of fire.

"And your hot comb, Miss Kety?"

"Oh, honey, that's not used anymore because it damages

the hair. It doesn't last. The calm weather curls your hair back up into snails and you're just sitting pretty at midnight with your hair curling itself up." She laughs; Miss Kety and the neophyte laugh together while the sun bleeds over the neighborhood that now boasts a new Center for Psychiatric Treatment next door. The expert hand now rests on Yetsaida's broken nose.

"Ah, you too."

"Yeah." And Miss Kety feels something like a tickle from the past run along the map of the world on her forearm. The beautician falls silent. She begins to massage the head with her hands, so the cream will penetrate properly, and the mechanical action is a relief for her.

"Sometimes you have to remake yourself because the first time you came out badly. You have to give a hand to destiny, use and buy the best Swedish Biolage products, they aren't cruel to animals or to the environment."

You have to sit down and look at the sunset with your eyes half closed, in the beauty parlor high above the Trastalleres projects, under the hands of Miss Kety. Miss Kety also closes her eyes, and she remembers a hymn that she learned the other day at the Church of the Latter Days, where she goes whenever she has the money for the tithe. She continues the ritual in silence. Under Miss Kety's hands, the neophyte has become pensive. She dreams of her new hair, dyed Auburn Copper red and floating in the breeze as she appears before the Sky Academy for Looks and Beauty in Miami, recounting in perfect English the trajectory of her success (*I was a small girl with a dream, a dream of beauty that has come true. Thank you, Sky Academy!*). Miss Kety looks at

Yetsaida's broken nose, and with one eye closed looks at her own, and she raises the girl from her seat before the sunset.

"I have to give you a conditioner so you don't walk the streets with so many split ends." And she walks her over to rinse out her hair and give her a shampoo.

Abnel, Sweet Nightmare

Finally, goddammit! It's about time! she thought, so loudly that she would have sworn the man sitting next to her had heard. She had been waiting almost 45 minutes for the bus. Seeing it in the distance, she automatically pulled a quarter from her purse and prepared herself for the upcoming simulated war: force herself onto the bus as best she could, as though there were lots of space, and then fight to the death for the last open seat. She was an expert at this. Years of practice had strengthened her knees and elbows, made her torso and waist agile. She knew all the most effective contortions to arrive at The Seat. She practiced twice a day, and thus had arrived at perfection: finding herself the happy owner of a non-upholstered seat by the rear window of the bus. Freedom from the overwhelming shower of dust, from the quick handling and the casual friction at every stop, and from the undeodorized armpits of all the others who, just like her, practiced the bus ritual every afternoon at five o'clock. She situated herself on her square of plastic, rested her purse on her lap and held onto it with both hands, just in case. Habit led her to make a quick review of her companions: a couple of students letting their thighs touch obliquely, a girl roughly 20 with two little kids, a man who was asking for a Vinyater Street in some development with a name from a Spanish saga. *Mary, mother of Christ*, she thought, bewildered.

She calculated the time. It was already late, but if the bus ran the light, if it turned onto Loiza Avenue quickly, and if the driver didn't stop at every bus stop, she would get there on time. She would run up the stairs, throw open the door, and abandon her body to the ecstasy of seeing him. *Him*, she was crazy about him, his ebony black hair, his nervous walk, seeing him step wet and wrapped only in a towel from the bathroom. Abnel Nieves—it's written on his mailbox, the name plate by the intercom says it, the curve of his back says it, the air he cuts with his profile says it, and the drops of water sliding down his thighs—Abnel Nieves.

A sharp curve jolted her from the thoughts that were warming her insides. She was shocked by the light sweat that glistened on her chest. She passed her gaze over the intestinal walls of the bus—*Christ loves you, Grab your neighbor's ass, Free the political prisoners, Carmen and Caco Forever*. Carmen and Caco, Romeo and Juliet, Abnel and her forever, for always. And she thought of how every weekday at six-fifteen sharp she strolled to the window of her efficiency apartment and gazed from behind the thin curtain towards the building across the way to see him step wet from his bathroom, select his pants and shirt from the closet, dress slowly and go out to who knows where. Occasionally, he turned towards the window, and he granted her a view of his tangle of pubic hair, including even his tender penis, so that she would have nightmares that night. She always dreamed the same thing: she, the ugly librarian, curveless, skinny, was rescued by Abnel in a towel, who carried her from the bookshelves to his bedroom. There he kissed her nipples delicately; he passed his tongue across her belly and into her bellybutton, gently, gently. He kissed her and suddenly he transformed into a beast who wanted to take her at any cost, eat her alive and leave her broken, in pain, for having dared to dream of his tenderness, a woman so ugly, so foolish and insipid. Seeing him naked, drying his body and then dressing himself without haste was her consolation. She couldn't miss Abnel this afternoon. He pulled her, even if it was an illusion, from her boredom.

It's five-fifty. The bus turns quickly onto Loiza Avenue; everything seems to be going fine. But suddenly the vehicle runs straight up against a tremendous traffic jam. Three minutes pass ... four. The bus doesn't move. It's locked in place like a putrid cyst. *God, oh God, why have you forsaken me?*

Five after six. What will happen to her if the bus doesn't arrive on time? How will she fill the empty hours of her afternoon and evening? What will she dream of? Abnel will get angry with her and close his window forever. He'll cut her nightmares out by the roots and leave her even more lonely. *It's ten after six—life taunts me.* Life taunts her for being old, for being ugly, for being flat-chested; all of this she knows well. Life wants to strip this last pleasure from her. She's about to cry; *Mary, mother of Christ ...*

Then suddenly, a miracle. At exactly six-fifteen the traffic begins to move, slowly at first. The Loiza is suddenly empty, giving hope to the hopeless. She's going to make it, a few minutes late, but she's going to arrive. Maybe she'll make it only for the streaking of Abnel's wet buttocks towards the bedroom, maybe only a rapid glimpse of his towel abandoned in passing, but it doesn't matter. Two more stoplights and she'll be at the corner by her house. "Get a move on, driver. Run the light, move it!" She pleads with her lips, her eyes pushing out of their sockets, waiting for permission to bolt towards the window. "Move it, driver, one more light, one more stop!" And the driver lets himself be guided by this hand he doesn't see, that of her will, stronger than steel, stronger than a city traffic jam, stronger than the time that is measured by alarm clocks and watches.

Six-twenty-five. The bus door opens at her stop. She flings herself out and runs wildly up the stairs. She throws her bags on the floor; her purse scatters its innards across the floor in relief. She rushes, disheveled, toward her window, late, late, ten minutes late. She forgets to keep herself hidden behind the curtain, almost tears it down as she grabs at it, hurls herself face first towards her window to try to recover, even if it's only a streak of Abnel. Panting, tearful, she stops dead in her tracks.

On the other side of the chasm, in the apartment across the way, Abnel Nieves is naked, looking at his watch, standing in front of the open window. Slowly he lifts his eyes from his watch to the hole where his neighbor appears. Abnel looks at her and smiles maliciously, gesturing toward his watch, silently forming the words "You're late." Then, smiling, he proceeds with his ritual, walking slowly towards his bedroom to dress carefully and deliberately for her.

Night Stand

There she goes, the proud girl, proud and bored with her breasts and her behind. Oh, she had them dancing that night, she had walked them up and down all the streets of San Juan: her new high heels, and her new hair, her new eyes and her sex new and smelling like something low, like the beach, something from the salt water. How she had laughed finding herself mirrored two and three times in the shop windows, the newest mannequins looking so much like her. How the San Juan boys gazed at her; and oh, how they'd return to their houses to pull on it in her name, how a few wouldn't wait for the bonanza of the bathroom, and how in the car parked on a dark corner they'd do it seven times seven, thinking of her breasts and her dancing ass and her sex sprinkled with neon, autographed by Lycra, over and over in so many shiny cars and shop windows, in so many rear view mirrors, in so many glassy and salivating surfaces.

There she goes. Again climbing the stairs of the dance

club and finding a corner near the bar, to sit down, to cross, no uncross, no cross her legs because she is thirsty. All night long so much looking at people sweating and twisting. Yes, she thirsts and hungers for justice. She wants a drink ... a seductive mouth on the other side of the drink, a seductive pocket full of lots of Ben Franklins to pay for the drink, and, pinnacle of the seduction, inside the pocket, a seductive key ring with three wondrous keys—the one that opens the door of an apartment in the Condado, the one that opens the door of a prestigious office, the one that opens the door of a Volvo and not a Subaru, Volvo and not a Subaru, Volvo and not a Subaru. She thirsts and hungers for justice.

Now she sits. Now she crosses and uncrosses her legs; now she tires of watching the couples wriggle on the dance floor. She is about to go when a hand sporting a gold watch and well-maintained cuticles (she readily recognizes them; she is a beautician) approaches. Along with the hand is the sleeve of an expensive jacket and down below, two fine creases of pants made from Italian linen, the finest expression of European shoes, a silk shirt, a tie, and a mustache that seems made of silk: *the man*. He draws his seductive mouth nearer ... he offers her a drink (justice), gin and tonic; he smiles and speaks elegantly. He asks her name—Stephanie—and he tells her about his stressful job as an industrial engineer in a stressful office decorated with stressful art deco furniture. She counts the first key—a prestigious office—and starts to let herself be seduced.

"Ay, what purgatory!" he tells her. "Papers, documents, work, always more work." For relief he goes to the sauna, lifts weights, and takes curative vacations in New York and San

Diego. She stares at him, she hears him, she listens to him, she lives him. She cries over the bit about the stress, she believes, she shares his desperation. She wants to always, always accompany him—standing next to his European shoes, his silk ties, his wallet, his wondrous keys. Hunger and thirst for justice.

The club closes. He keeps talking. He touches her bottom, elegantly, and offers to take her home. The breasts and buttocks of the girl dance jubilantly. In the car—second marvelous key, a Volvo and not a Subaru, Volvo and not a Subaru—he tells her it's nearby. And the apartment—third miraculous key, oh Condado, how her backside bounces—he tells her to make herself comfortable, he asks if she'd like a drink, he'll bring it to her, so elegantly, and he kisses her, elegantly, he caresses her, he falls over her like a mighty wave, he falls upon her and doesn't let her go.

A few minutes have gone by. Tangled up in the living room, with nylons and wrinkled pants down around their ankles, he pulls out a fourth key (the uncounted one) and quickly puts a condom on. While he jumps on her, gropes at her and messes up her hair, she dances ecstatically thinking of other things. She'll remember the street and the number. She'll ride around in his Volvo and he'll quench her thirst at an enormous price without glancing at the bill. Yes, and why not? Together they'll relax in New York and San Diego. She'll see them, all the others who wait in ambush for just this, and they'll see her. She'll see them stewing in their juices, drooling all over themselves. Over their shoulders they'll see her dancing breasts and behind, her new high heels, her new dress, with her new man, her sex all sprinkled, crossing, no uncrossing, no crossing her legs, but now not from thirst or hunger, not for justice, but—ah! what ecstasy!—from perfection.

Resins for Aurelia

> *Aurelia, Aurelia dile al conde que suba dile al*
> *conde que suba que suba suba por la ventana*
> Aurelia, Aurelia tell the Count to come up
> tell the Count to come up to come up come
> up through the window
> —Lyrics of a popular "Bomba"

Nobody knows how that fad started, but in less than three months all the whores in Patagonia had slaves on their ankles. You could see them walking around the town plaza on their day off, eating fruit and nut-flavored ice cream, or on the streets near the river. You could see them buying groceries at the marketplace with that little chain sparkling in the distance, the secret signal that disclosed their occupation. The sparkle down there on the ankle, the conspicuous way of calling it a "slave" would light up eyes and produce frowns all up and down the Humacao. In the direction of the river went the prostitutes' feet wearing slaves, and in the direction of the little chain went the eyes of all the women and men in town. Lucas, for one, under his wide straw hat, pruning scissors in hand, would stop feeding manure to the shade trees to see them go by, with hunger in his eyes.

His grandmother had taught him the craft of caring for flowers. Nana Poubart brought him as a child from Nevis to that grey town with a river that would devour the trees on the plaza. He didn't remember anything from his native land, only the concave terrain in his grandmother's breasts, the woman whose tongue got tangled up in sharper inflections than the rest of the inhabitants of the town; r's and t's a little more acute, more difficult to decipher among the sounds that crowded the air in Patagonia. The air in Patagonia—usually stinking of the swelling river, of humid matresses, of pee—stopped where the grandmother's air began, full of her concoctions of plants and flowers. Their house, although it was humble and flanked by the huts that served as brothels, always smelled like the resins of shade trees. The wood floors were shined with an amber-colored cream made out of the capá tree with beeswax and the essence of jasmine flowers. Right in front of the Conde Rojo bar, Nana had planted and braided a lemon tree and a guava tree. From the moment she planted the trees she nurtured them with apt fertilizers for gentle and abundant growth: whore shit mixed with menstrual blood. Lucas was ashamed when Nana would send him to the back door of the Conde Rojo to ask the madams for their pupils' basins. He protested with his feet and with his chest, but Nana would have nothing to do with badmouthing, nor with gossip inspired in false modesty. According to Nana, there was nothing better to grow shade trees, nor medium-sized fruit trees, on this side of the Caribbean.

That is how Lucas became used to whores; to their smells; to their textures; to their looks of complicity. He slept with them from prepubescence, starting at the age of twelve. Under the pretext of giving him their basins full of shit, the madams and the older whores would make him come into the Conde. There they would force him to wait while they changed their clothes, powdered their fallen or full breasts with perfumed talc and multi-colored powder puffs made of foamy cotton. Sometimes they would commit him to the task of tying the garters that held up their stockings, or to undoing the buttons of their corsets. And after these furtive frictions, some of them would French-kiss Lucas on the mouth, performing maternal affections on him, shitting amorously before him in the basins, and guiding him once again back to the entrance of the huts.

Meanwhile, Nana would wait for him, sitting in the mahogany and straw rocking chair on the balcony of the little house. She had braided the guava tree in the entrance herself, with her ironing hands, hands that washed rich people's clothes in the river. She taught Lucas how one takes the branches of young trees to make designs on their trunks. "The fingers," she would say while she spread whore shit on them, adding the resins from rubber trees and honey, "it is important to know where to place the fingers and how much pressure to apply to bend the tender cortex of the trees without breaking them." Year after year, Nana sensitized his fingertips to the extent that Lucas learned to take the pulse of trees, the shade ones, the fruit ones, and the flower ones. He could feel the sap running through their veins, and through careful measurements of temperatures and the pressures of their liquids, he knew if they needed water, pruning, or a bleeding to release excess resins from their interiors.

What Lucas could never get used to was to the pungent smell of whore shit. Even though he continued going to get the basins each time Nana sent him, and continued sleeping with them, he could never sink his hands into the basin with good cheer. He convinced his grandmother to let him try other methods, and he set upon the task of scouting the river banks with a machete and a coffee can, bleeding the sap of all the bushes and tree trunks and plants on the littoral.

Nana also knew how to extract the spirit from plants, how to use the leaves to cure love sickness, colic, diarrhea, vomiting, brothel fevers, and other aches and pains that troubled her neighbors from Patagonia. She knew about teas against the pain of menstrual cramps, and about orange infusions to quiet down weeping and tremors, soursop leaves to alleviate bloating, cataplasms made with resin of the hog plum plant to revive the skin's heat. She knew a million of these secrets. And in the same way that she mended roots and trunks and foliage, she also mended bones, broken vertebrae, women's eyes that had "accidentally" run into doors, purplish bruises, blood clots, sprained ankles, miscarriages and abortions. It was healing people that supported Nana and her grandson. But Lucas didn't find what she did with her plants and her hands in the service of people that interesting. The people smelled like shit; they gave him only a fortuitous pleasure that left him lonely, melancholy, and confused right after the last tremor. Not trees. They had their thickness and their richness; the soft humid green of the avocado leaves, the skin of *palo santo*, or the little cortexes of spurge provoked the sweat of relief on his skin. They would leave him calm and clear. What he enjoyed the most was taking the resins out of trees, making them bleed deep, gummy ambers with which he was sure he could mend anything: the bones that Nana fixed, the trunks of flowering guava trees, after-pains of the soul; delicate ointments to weatherproof wood, avoid leaks and humidity stains on roofs, shape table legs, make a body breathe. The resins could do it all.

When Nana retired, she dedicated herself completely to healing damaged whores, and Lucas, already of age, was hired as the municipal gardener. Nobody had ever seen vegetation grow with such beauty under human hands. Lucas, the son of the washerwoman from the islands, transformed the naked plaza of a salinized town into a divine garden, where pansies would grow in the sunlight, *duendes* and *cohitres* cohabited without wilting under fruit trees, pink and yellow oak trees stood up tall in the direction of an eternally grey sky. Now it was adorned with a paradise of plants and an elegance that he had made. All the high-society ladies would give him work in their homes; he would perform beauties in their interior patios, in the entrance walks; he would plant and take care of royal palm trees, coconut trees; he was able to combine azaleas with gardenias with rosebushes and different colors of hibiscus; he could braid spiny bougainvillea so they would hang their manes over the terraces and the roof tops; he could fill the house with the smell of his ointments for mahogany tables and for old slanting roofs; he would shine the floor with the amber resins of thousands of trees that he distilled in the back rooms of his little house in Patagonia. He arrived and left everything smooth, fresh to the touch, slippery; he would protect surfaces from the grey ocean spray that covered the town with crystallized vapors, and he would smooth the wrinkles of time, giving back the secret palpitations to each trunk or torso that had the opportunity to receive the gift of his fingers. Lucas' fingers. Some circumspect women had surprised themselves dreaming of Lucas' fingers; that he would pull from deep inside them all the dryness that had been so well stored up inside, that he would undo them in rivers of succulent amber, dense musks smelling of deep and secret fragances, those from which they protected themselves to preserve their respectability.

And it was strange how people treated Lucas, because no one except Nana and the little Patagonian whores could look him in the face or let their eyes slide down the rest of his body. Almost no one held his stare, almost no one remarked his features, the dark and sweet almonds that were his eyes, nor how wide and remote his smile was. No one, except the whores, noticed how wide his back was, fibrous like ausubo wood; nor the perfect roundness of his mounds of flesh, there, above the thighs, nor the deep mahogany color of his skin, always fresh like the shade. And nobody even dared to slant the tail end of a glance down to the extremity of the root that announced itself succulently between his pant legs; the wide knot that promised trunks of dark and succulent flesh, soft little hairs smelling of sea grapes. He didn't even notice how beautiful he was, because like everyone else, his attention was fixed on the precision of his hands. His fingers, long like a bird's, ended in curved points, with diluted crescents at the bottom of each nail. These were always bordered by fragments of earth, and grooved sometimes by the very fine fibers of keratin that created different and masterful textures on each one. The palms were wide, fleshy, with calluses on each finger. Deep veins and subtle cuts furrowed them on both sides, forming little destiny maps on the whole ripe acerola-colored surface. But, surprisingly, Lucas' hands were soft in their strength and precision; shy and soft like when he was a child and carried basins of shit to his grandmother's house; shy, soft, and flighty in their strong pressure on things. All eyes that stumbled on Lucas looked at his hands, just like they only focused on the little gold chains clinking on the ankles of whores.

The first time the river flooded the gardens that Lucas had been knitting throughout the town plaza, it pulled up a congregation of plants at the root. It had taken the gardener almost four years to build his vegetable empire. Lucas had just pruned the cedar trees and the rubber trees, curing them of parasites and other tropical diseases that afflicted them. The buckets of resin filled with mud; the currents undid the tourniquets straightening trunks the winds had left crooked. But he had known this was bound to happen sooner or later. He had known it ever since he started scouting the river banks in search of resins and noticed that the river bed was artificial: it had been detoured deliberately to comply with the expanding needs of the municipality. "Things have their lives and their deaths and follow their course on earth. That cannot be changed by the hands of man," Nana had said to him when he told her about his discovery. And the words uttered by the old healer were providential, because weeks later the river took it upon itself to recover its original course and flooded the town. The biggest loss was not the town gardens. Due to the unfortunate whim of the Humacao, more than two hundred people died, almost all of them from Patagonia. Among them, Nana.

It was an affair of destiny. After work, and after distilling two gallons of Tabonuco resin in the back rooms of Nana's little house, he went to fetch the whore shit from the huts. One of the girls, honey-gold like the substance he had just distilled from the heart of the trees, opened the door, her eyes, and Lucas' heart. She was new, he hadn't seen her before, but that afternoon she offered herself to him for twenty dollars, and he left thirty on the dressing table made of pine slabs, next to the little mattress where they made love until dawn. The rattling of the rain could be heard in the distance while he penetrated her softly during the first round of caresses and she opened silently to let him come in between her legs. Lucas spent time on top of her, moving like the willows in the cemetery. He noticed that at first the girl was only doing her job, but little by little the hinges in between her legs started to moisten, smelling like a newly cut cedar tree. Then Lucas moved more hastily until she arched her little back in a thrust, he stuck his ribcage to her chest and released in a languid and sad sigh, while her vulva throbbed around him inside her. Three, four times she dissolved under him. When she was exhausted and had forgotten herself, and while the heavy rain threatened to dissolve the roof slats at the Conde Rojo and the river roared and dragged away in its roaring course half of the inhabitants of Patagonia, Lucas Poubart penetrated the woman for the fifth time. With the first push he felt that his stomach was filling with all the juices his body had been capable of producing in the years he had existed on the face of the earth. And he emptied himself completely into that little golden woman, while she covered her face with her hair, trying to keep him from seeing the face of death in the midst of the disaster that was their passion.

Fate saved them both. They had passed the river's swelling in the highest of the huts that made up the brothel. But the rest of Patagonia was pure desolation. It was located on a steep slope, close to the river. The waters of the Humacao had reached the plaza and, what was worse, trapped Nana in her bedroom, where her neighbors found her, dead. When Lucas arrived, he found the neighbors untangling Nana's corpse from the sheets that had tied her to the pillars of the bed. With one deep scream he dissolved in tears, embracing his grandmother's dead body.

It was close to noon when Lucas finally came out of his stupefaction, let go of Nana's body over the kitchen counter, and went out on the street to help the rest of the people. With water up to his waist, he saw people trapped amid the debris, boards, branches, and mattresses from houses that had been destroyed by the current. Thinking of Nana and what he had learned from her, he began disentangling the dead, saving those who were still alive, taking the mud from out of their noses and massaging their submerged lungs. He gave artificial respiration, warmed body parts, embraced orphans and widows. He took them to higher ground, out of danger, and by nighttime he collapsed from exhaustion on one of the benches in the shelter that the municipality had opened for victims of the disaster. He slept there without moving all night.

When Lucas awoke, he found that the river had receded to its original level. He went back to his house to make arrangements for Nana's burial. He didn't call a funeral home, but rather, went to the little room for distilling sap and cleared a place on his work table for Nana's already stiff corpse. From the flooded workshop, he rescued a can that, miraculously, had not been taken by the swelling of the river. Inside the can was a heavy ointment with a pungent smell. He opened the can, rubbed the balm on his hands, took Nana's clothes off, and with that cataplasm massaged her swollen and grey body. It took hours: face, jaw, neck, ears, hair, and then lowering his fingers and pressing against the shoulders, against the strong arms of that woman who had raised him since childhood. He took her fingers, so similar to his own, he filled them with the distilled ointment, he moistened them with his own silent tears. He smeared her chest, being careful to apply less solution on the dark aureoles. He lowered his hands, pushing hard on her belly and then her legs. He opened Nana's legs, caressed her grey pubis, and tenderly filled the cracks with sap—expert, connoisseur and humble in his duty of returning the smoothness and the moisture to his grandmother's dead body. He put it under a warm light, and waited three hours. Later, he wrapped her with a dress he had bought for her days before and went to the patio to finish making a coffin of polished mahogany, lightly stained a reddish brown that matched Nana's skin perfectly.

Four days after the burial, which was the most beautiful of all the burials celebrated in Patagonia, he went to what was left of the whores' huts to look for the golden woman. He didn't find her. Nobody could tell him for certain where she was. Doña Luba, one of the oldest whores in the neighborhood, told him the rumors that her father had come down all the way from Yabucoa to take her away. "That damned man was the first one to disgrace her. Aurelia herself told me when she had just arrived in the *barrio*. When she heard he had found her, she took advantage of the chaos of the flood and ran away. She must be hiding somewhere. If you see her before I do, tell her to leave the Conde Rojo and come work for me. If I see her first, I'll tell her you're looking for her."

While he waited for news about Aurelia, Lucas concentrated on repairing the gardens in the plaza. One day they called him to the mayor's office. There they told him that they required his services, but for a different occupation than reviving plants in the plaza. There were still bodies floating in the river that the current had dragged to the outskirts of town, bodies that nobody wanted to go pick up and that were fostering disease. "They are the corpses of whores. Nobody wants to touch them. We fear the worst: epidemics, plagues, pollution. We can't take the risk of having those bodies dragged by the current to neighboring towns. A scandal like that would soil the good name of the mayor." Lucas accepted the task. He asked for transport to scout the river banks, on the condition that they raise his salary and grant him total independence in selecting the plants and trees to be seeded in the town plaza.

That was how Lucas, the municipal gardener, became the retriever of drowned whores' corpses. Much to his surprise, dead prostitutes kept showing up in the river waters long after he had removed all those that had drowned during the flood. Once in a while he was called in by the municipality to go pick up stray corpses. "Another whore drowned by the flood," the policemen who called Lucas to work said with muffled laughter. He got used to the custom after the first months and would go out on his own, patrolling the river banks to save the officials a call and avoid interruptions to his routine as a gardener, to which he returned after the first round of rescues.

It was always the same story with the recovered corpses. First he would jump in the river, swimming, to disentangle the bodies from the plants that stopped drowned whores from floating downstream. He would disentangle their hair to see if he could identify them. When he got to them, some already had their lips pitted by the fishes, or their eyelids populated by crustaceans, and their insides inhabited by little shrimp and water fleas. It was hard to identify them, if it wasn't for the little chain on the left ankle that revealed their profession. Those that were disfigured he carried softly, as if they were asleep, and deposited at the morgue. With those whose death was fresher, he would become attached, without knowing why. Then, he would take them home. He would prepare some sort of oil for them with a pleasant essence to take from their faces the rictus of surprise at finding themselves drowned, the nightmarish fear on their faces and in their muscles. He would expertly caress their flesh, smooth their faces with his hands, thinking about how nobody was going to claim them, how they would be thrown in the garbage dump, cremated, without a single good-bye caress. The whole town had touched these bodies and now wanted nothing to do with them. "Nobody wants to touch you," Lucas would tell their nether parts, "nobody wants to touch you, and nobody would know how to do it better than me." He didn't do much for them, he knew. But when he delivered one of those new bodies that had provoked his affection to the morgue, he took pride in how beautiful they looked, with their skin smooth and oiled, smelling of fresh mint plants. Before taking them on his municipal vehicle again, he would unfasten the infamous little gold chain from their ankle, and he would keep it in the pocket of his pants. Maybe that way they would be treated better.

One day Lucas was walking along the banks of the Humacao, not thinking about anything in particular. It had been so long since he had looked for resins or rescued any corpses. It was always the planting of gardens and spreading resins on roofs, tables, and chairs in rich people's houses. He stopped against a *caimito* tree to look at the veins of the trunk and to caress them softly. It was then that he noticed a pile of clothing that jutted out from the high grass on the other side of the water. His vision sharpened; it looked like a corpse. Cheerfully he took off his shirt, jumped into the river, and waded calmly through the shallow water. Approaching the pile, he saw small hands with girlish fingers that showed a hint of amber on the wrinkled, grey skin. This was a recent death, no more than a few hours, a night and its dawn in the water. Barefoot, with fingernails painted red, totally relaxed; and on the left ankle the infamous little chain. The flesh was visible through the blouse, letting Lucas see dark brown nipples that he thought he recognized. Carrying the corpse, he reached the other shore and began disentangling debris to see the dead woman's face. But as soon as he took her out of the water and into the sunlight, placing one of his broad hands on her drowned face, his skin felt the shock. It was her, finally her, Aurelia. He had found her.

But she was dead. Lucas wanted to cry—he couldn't. Eight months had passed since the terrible flood. The only thing left of that woman was his memory of a touch: awakening; feverish; new to him. He felt relieved, seeing himself freed from the spectre of that smoothness that set itself to his skin and left him nothing to do but long for Aurelia. He thought that now he would go back to being the same man, the man who would have never abandoned Nana on a rainy night, the one who could graft trees and make himself desirable to the other prostitutes. Maybe he could even find a good woman to live with in his grandmother's little house, and he would become the man Nana had raised, redeeming her thus from a useless death. Then he deposited Aurelia in the bed of the pickup and drove toward the town.

He took her home and undressed her. He took off the pieces of cotton blouse, the red panties and the broken skirt. He took off the little gold chain, which he threw in with the others in a pewter cup he had bought for that purpose. He took out the combs and began by untangling the dense mat of hair that he had held in his fingers the whole night of the flood. As soon as he buried the comb in her hair, little bugs started coming out, which he killed with the tips of his fingers: river spiders, fleas, larvae of insects that had gotten caught. He combed her hair softly and with a smile on his face, continuing until the hair was completely smooth. He washed it with soap and showered it with rose water. He sat in a rocking chair next to the fresh and humid corpse of the golden girl while her hair dried. Then, still smiling, he walked to his resin workshop and took out the can he had used almost a year ago to prepare Nana for her tomb. There was enough solution inside to cover the body, small as a bird, that lay on the table. He had never used it on another—instinctively he had kept what was left, maybe for this woman.

He began the ritual of smearing his hands with the solution. He started at her feet, toe by toe, ankles freed of chains, rigid legs, all of her oiled by the resin which, having aged, gave off a subtle smell of woods of all sorts and condensed flowers, a vegetable smell from which none of its original components could be distinguished; smelling a little like shit, a little like blood. Pressuring attentively, he relaxed all the dead woman's muscles until he felt that the friction and something else returned warmth to her skin. With that sensation of strange temperatures between his fingers he continued his way up Aurelia's body. He spent forty-five minutes massaging her strong, caramel thighs, whose tawny hairs reflected the light of the workshop. And there again he felt the strange heat that returned, from the inside out, to the girl's body. Lucas saw how from the inside of her thighs emanated delicate droplets of water, a sweat that didn't smell human, but rather like river banks. He continued the massage, putting his hands under the girl's legs, pressing her buttocks, which also lit up under his resinous fingers. He felt a pounding of hot blood between his legs; he looked at his erection, aching from the desire of rubbing completely against her on the workshop table.

Lucas shook his head, paused to see how from the waist down, his little drowned whore had recovered some color and emanated vegetable smells from the pores that expelled water. He smeared his hands again, this time placing them precisely on the dead woman's face. He made circles on her forehead with his fingers, on the cheekbones, the eyelids that he opened and closed, to let her rest from his fingers' pressure. The lips, the jaw, the spine, and the nape, he put back in their place. The shoulders and collarbones relaxed under the pressure of the gardener's fingers. He put her on her side to apply resin to her back up to the buttocks, already hot, that perspired onto the wood of his worktable, against the palms of his hands, against the desires which grew despite his concentration. He turned her over again to apply resin to her adolescent breasts. The resin's heat made them release the river water that they had sucked in while she drifted. Her nipples, hard and dark, took on magical colors, and Lucas could stand it no more. He took off his clothes, put a little bit of resin on his pelvis, on his pubis, and on his penis. As he opened the drowned woman's legs he felt the stinging heat of that viscous unguent, felt himself burn. With his fingers he loosened her vulva, and right there, on the workshop table of grafts and woods, he penetrated sweet Aurelia, Aurelia of ambers and resins, his beloved little whore, to be able at last, at last to fill her with heat. Death was simply a turn of fate. His hands could not scare it away. But his dick and his resin, that burning heat that came back, wrapped in vegetable consistencies—that was present, the product of his hands and his longing, the insistent memories fastened tightly to his fingers, to his skin.

He came inside her, contracting all his back muscles, he emptied into her like a milky placenta between her legs, screaming into her ear that he loved her, that he loved her as she was and forever. He fell asleep on the corpse and dreamt that the golden girl enveloped him in her arms and gave him small kisses of love.

When he awoke, Lucas went to the cup full of chains, took hers out, and put it on her ankle again. He put the body under a warm light, went to town, and came back with two big blocks of ice, a knife, and brass buckets of the kind he used to collect resins. He used the opportunity to tell the municipality to find someone else to rescue drowned whores. Then he went back to his gardens, to his walks in search of resin, and to his excursions (less and less frequent) to the prostitutes' huts in Patagonia. Three times a week he would lock himself up in the workshop of the maternal household with a can full of unguents and a bottle of rose water, and would not come out until dawn, smiling and covered all over with sticky sweat.

Urban Oracles

The street wakes up early. Buses, cars, a woman collecting the first abandoned cans of the morning, fresh—still holding liquor and cigarette butts from the night before. People drink at all hours. The street awakens, like all salaried workers, early every day, and all the creatures of the street awaken at that hour. They unshuffle their dreams to see if they become truths. Truths come out in dreams, the disguised designs of coincidence. Mama always said so, and in her practice she was outstanding at reading dreams, reading accidents and reading the steps of the events in the street. "The street is an intestine," she told me; "if you look carefully at the marks, you can forecast the future."

All the street creatures wake up at the same time, air creatures and land creatures, underground creatures, creatures that live on carrion and garbage. The pigeons wake up at exactly the same time as the street. And the salaried workers wake with the street. Still surprised that the world is moving, they wake up distressed by the dreams of future accidents which they have already dreamed so many times. They awaken expecting attacks, failed negotiations, matrimonial troubles, traffic jams.

Mama used to take me on Sundays to the Plaza de Armas to watch the people go by. In her line of work this was an indispensable exercise.

"If you want to follow me in this career," she would tell me, "you have to learn to read people. The guardian angels reveal twenty percent of what happens, but the rest is interpretation. You must learn to read people. Watch the way they walk; try to smell them as they pass. Note if they have a weepy or a lost look in their eyes. That one, look at the way she's walking. Either her feet are sore or she doesn't want to arrive at her destination. She's a candidate for a consultation. And that one with such a careful hairdo, so well-dressed, but a bit too tight. He's also a good candidate for a consultation."

The street wakes up. The pigeons wake up and fly away from the abandoned employees' club of the Banco Popular. Now I must, just like every Wednesday, stand here and watch the people pass by, study them and sniff their glances.

It's early when the young woman climbs down from the van full of people. She wants to make all the inquiries today. She wants to get through the agony quickly. Her husband issued an ultimatum. He wants $15,000 to cover his part of the mortgage on the house that they bought when they were together, and then it will be completely hers when the divorce is final. If she doesn't get the money soon she'll have to move. But where? Move back to her parents' house? Begin again? The woman climbs down from the van in front of the Government Employees' Union. She doesn't know if they'll give her the loan, she doesn't know how twisted up her finances are, she doesn't know anything at all. The tears almost come pouring down. Her heart almost jumps from her chest. She almost opens the door to the building when suddenly—splat—she feels something heavy fall on her, hit her on the head.

What's this? What was that? What happened? I raise my hand to the crest of my head and it comes away covered with blood, and I think that a piece of soffit, a cornice, has split my head. My ears are ringing, there's a vacuum in my head, it's almost a relief. I touch it again and more blood, but this time feathers too. And feathers. I pull feathers from my hair. I don't understand anything, and there's that buzzing from the shock in my head. The people crowd around me. The ticket taker points at the ground. They all look down. There's a dead pigeon with a broken neck, stretched out in a pool of blood. Everyone looks at the pigeon. I don't know where to look. It disgusts me, it disgusts me completely. I almost tear my hair out to get those feathers off of me. I shake myself, the people brush me off.

"Buy her a coffee, a little coffee ... Honey, it ain't nothin'. In a few minutes you'll go on home ..."

"What home?" I'm about to scream. I shake my hair out; it comes out in bunches. The vacuum buzzing in my head, and the people far far away, as though I were in a tunnel.

A fortyish woman touches me. I spring around. She stares

fixedly into my eyes with a calm that tranquilizes me. She takes my hands and in the right one she leaves a card. I hear her tell me slowly,

"I don't know what your religious beliefs are, but these things are often divine messages. Maybe somebody wants to tell you something from on high." The woman smiles at me and disappears into the crowd.

The street awakens at the same time everywhere, and everywhere there are creatures roaming aimlessly along it, consumed by loneliness and sadness. They don't have anyone to talk to. They wake up and don't have anyone to touch, or they touch someone who has already gone out into the street, or they don't have time to even realize that they need to touch somebody, which is the worst. The street wakes up and turns everyone into zombies. Each one remains in his own street of dreams and accidents. They cross in front of each other to eat, to punch the time clock upon arrival and departure, or to effect transactions and buy things for themselves that soon enough they will throw into the garbage can. And then the street reclaims all these purchases, these dealings, from the garbage and throws them back again into the faces of its creatures. The street is merciless, and it awakens everywhere at the same time.

Mama said that the powers of the street are inherited, that she would pass spiritual clarity to me in my blood so I could earn my living that way, by communicating with the spirits of the Light. She worked with the Benevolent Forces.

She read people's auras. She could read whatever they had touched, coffee grounds laying in their cups, their palms, or cards. For her, everything was an open book. She promised that I would inherit the powers. She also showed me the art of calming people with a certain tone of voice and a specific hand temperature.

"Dry and lukewarm," she told me. "They must always be dry and lukewarm. Put on flower water, or some gentle violet essence. That calms people. Look into their eyes. Speak softly but firmly. Smile."

I walk to the corner of the hospital and study the fauna of the street.

"I don't know what I'll do," thinks the woman, "if the biopsy is positive." She was alone in her house when she felt the hard little ball in her left breast. She wasn't the type to go to the doctor often. She felt the nodule and didn't want to go to the hospital. It was very small; surely it was nothing. So she continued in her habits, continued in the routine of a retired teacher, the routine of a grandmother caring for her grandchildren, cleaning yards. She continued until she felt the piercing point of pain. And then suddenly she raced to the doctor, to the hospital, to the surgeon. All of it in a blind rush.

A note arrived advising her to pick up the results of the biopsy, but although she approaches the hospital every day, she can't gather enough courage to enter the laboratory which will determine her future.

"Let it be nothing, Holy Virgin. Let it be nothing." She loses her pace; she walks scowling down the street, nervous, praying out loud. Nobody and nothing around her matters to her. The cars don't bother her, nor the beggars asking for spare change. She doesn't care if she is sweating, nor if people see her go by every day on the sidewalk in front of the hospital. Nothing matters to her, not even her life. Fear has occupied her completely; it doesn't let her breathe. Fear nullifies her life. "My God, drive off this fear," murmurs the woman, and—WHAM!

"What is this? What was that? What happened?" I raise my hand to the crest of my head and it comes away covered with blood. I think, "Someone threw a stone, or a piece of soffit." My ears are ringing, there's a vacuum in my head; I lift my hand to my head again and more blood, and feathers as well. And feathers, I pull feathers from my hair. I don't understand anything, my head is buzzing from the shock.

"My God! ... God?" Everyone begins to crowd around me. I think I'm going to faint.

"Hold her! Hold her up!" someone screams. A woman opens a way among the multitude. She touches me; I spin around and she looks firmly and sweetly into my eyes. She takes both of my hands, gently. She smells of violets, of flower water ... I calm down. She puts something in my right hand as she says to me, "You could be Protestant or Catholic, but you can't ignore the designs of heaven."

She leaves. In my hand she has left a card: Lady Faustina, spiritual consultant. It's only then that I notice that at my feet, near the street, a dead pigeon is stretched out in a puddle of its own blood.

∽

People came to the house in torrents. Mama read them the past, the future; she prescribed prayers for them, and magnetic stones. She communicated with the Indian, with many sacred spirits. She heard the voices, mumbled secrets. But with me, no matter how hard I tried, nothing. I continued to be deaf to the spirits, daydreaming. I didn't see the lights, or hear the voices.

"Don't despair," she said to me. And at the meetings other young people were already communicating with protective ancestors. Mama told me, she promised me, that I would inherit her faculties. And she lied to me, she lied to me. So many dreams of glory, the heir of Lady Faustina, her daughter Tinita. She said to me, "You'll have more powers than I do."

I wanted to cure people, find lost objects, dedicate my life to the Lord in his works. I wanted to know those forces that move objects and events from below, from the deepest sources. Not the people with their faces of death, recently awakened, stumbling along the street. I killed myself studying and reading, memorizing books of prayers, meditating for hours. I studied the people; nobody read postures and glances better than me. Nobody was better at guessing their sorrows. But it was guessing, guessing. It wasn't a revelation. From Mama I inherited nothing more than dreams. What a trick, Holy Virgin! And like a consolation prize, this one ridiculous faculty ... this useless ability to break pigeon's necks in full flight.

Marina's Fragrance

Doña Marina Paris was a woman of many charms. At forty-nine her skin still breathed those fragrances which when she was young had left the men of her town captivated and searching for ways to lick her flanks to see if they tasted as good as they smelled. And every day they smelled of something different. At times, a delicate aroma of witch's oregano would drift out of the folds of her thighs; other days, she perfumed the air with masculine mahogany or with small wild lemons. But most of the time she exuded pure satisfaction.

From the time she was very little, Doña Marina had worked in the Pinchimoja take-out restaurant, an establishment opened in the growing town of Carolina by her father Esteban Paris. Previously, Esteban had been a virtuoso clarinetist, a road builder and a molasses sampler for the Victoria Sugarcane Plantation. His common-law wife, Edovina Vera, was the granddaughter of one Pancracia Hernandez, a Spanish shopkeeper fallen on hard times, for whom time had set a trap in the form of a black man from Canovanas. He showed her what it meant to really enjoy a man's company, after she had lost faith in almost everything, including God.

Marina grew up in the Pinchimoja. Mama Edovina, who gave birth to another sister every year, entrusted Marina with the restaurant and made her responsible for watching Maria, the half-crazy woman who helped Mama move the giant pots of rice and beans, the pots of *tinapa* in sauce, of chicken soup, roasted sweet potato, and salt cod with raisins, the specialty of the house. Her special task was to make sure that Maria didn't cook with coconut oil. Someone had to protect the restaurant's reputation and keep people from thinking that the owners were a crowd of sneaky blacks from Loiza.

From eight to thirteen years of age, Marina exuded spicy, salty, and sweet odors from all the hinges of her flesh. And since she was always enveloped in her fragrances, she didn't even notice that they were bewitching every man who passed close to her. Her pompous smile, her kinky curls hidden in braids or kerchiefs, her high cheekbones, and the scent of the day drew happiness from even the most decrepit sugarcane-cutter, from the skinniest road-worker burnt by the sun, from her father, the frustrated clarinetist, who rose from his stupor of alcohol and daydreams to stand near his Marina just to smell her as she passed by.

Eventually, the effect Marina had on men began to preoccupy Edovina. She was especially worried by the way she was able to stir Esteban from his alcoholic's chair. The rest of the time he sat prostrate from five o'clock every morning, after he finished buying sacks of rice and plantains from the supplier who drove by in his cart on the way down to the Nueva Esperanza market. Marina was thirteen, a dangerous age. So one day Edovina opened a bottle of Cristobal Colon Rum from Mayaguez, set it next to her partner's chair, and went to look for Marina in the kitchen, where she was peeling sweet potatoes and plantains.

"Today you begin working for the Velazquezes. They'll give you food, new clothes, and Doña Georgina's house is near the school." Edovina took Marina out the back of the Pinchimoja over towards Jose de Diego Street. They crossed behind Alberti's pharmacy to the house of Doña Georgina, a rich and pious white woman, whose passion for cassava stewed with shrimp was known throughout the town.

It was at about this time that Marina began to smell like the ocean. She would visit her parents every weekend. Esteban, a bit more pickled each time, reached the point where he no longer recognized her, for he became confused thinking that she would smell like the daily specials. When Marina arrived perfumed with the red snapper or shrimp that they ate regularly in the elegant mansion, her father took another drag from the bottle which rested by his chair and lost himself in memories of his passion for the clarinet. The Pinchimoja no longer attracted the people that it used to. It had lapsed into the category of breakfast joint; you could eat *funche* there, or corn fritters with white cheese, coffee and stew. The office workers and road-builders had moved to a different take-out restaurant with a new attraction that could replace the dark body of the thirteen-year-old redolent with flavorful odors—a jukebox which at lunch time played Felipe Rodriguez, Perez Prado, and Benny More's Big Band.

It was in the Velazquez house that Marina became aware of her remarkable capacity to harbor fragrances in her flesh.

She had to get up before five every morning so she could prepare the rice and beans and their accompaniments; this was the condition that the Velazquezes imposed in exchange for allowing her to attend the public school. One day, while she was thinking about the food that she had to prepare the next morning, she caught her body smelling like the menu. Her elbows smelled like fresh *recaillo*; her armpits smelled like garlic, onions and red pepper; her forearms like roasted sweet potato with butter; the space between her flowering breasts like pork loin fried in onions; and further down like grainy white rice, just the way her rice always came out.

From then on she imposed a regimen of drawing remembered scents from her body. The aromas of herbs came easily. Marjoram, pennyroyal and mint were her favorites. Once she felt satisfied with the results of these experiments, Marina began to experiment with emotional scents. One day she tried to imagine the fragrance of sadness. She thought long and hard of the day Mama Edovina sent her to live in the Velazquez household. She thought of Esteban, her father, sitting in his chair imagining what could have been his life as a clarinetist in the mambo bands or in Cesar Concepcion's combo. Immediately an odor of mangrove swamps and sweaty sheets, a smell somewhere between rancid and sweet, began to waft from her body. Then she worked on the smells of solitude and desire. Although she could draw those aromas from her own flesh, the exercise left her exhausted; it was too much work. Instead, she began to collect odors from her masters, from the neighbors near the Velazquez house, from the servants who lived in the little rooms off the courtyard of hens, and from the clothesline where Doña Georgina's son hung his underwear.

Marina didn't like Hipolito Velazquez, junior, at all. She had surprised him once in the bathroom pulling on his penis, which gave off an odor of oatmeal and sweet rust. This was the same smell (a bit more acid) which his underpants dispelled just before being washed. He was six years older than she, sickly and yellow, with emaciated legs and without even an ounce of a bottom. "Esculapio" she called him quietly when she saw him passing, smiling as always with those high cheekbones of a presumptuous Negress. The gossips around town recounted that the boy spent almost every night in the Tumbabrazos neighborhood, looking for mulatta girls upon whom he could "do the damage." He was enchanted by dark flesh. At times, he looked at her with a certain eagerness. Once he even insinuated that they should make love, but Marina turned him down. He looked so ugly to her, so weak and foolish, that just imagining Hipolito laying a finger upon her made her whole body begin to smell like rotten fish and she felt sick.

After a year and a half of living with the Velazquezes, Marina began to take note of the men around town. At Carolina's annual town fair that year, she met Eladio Salaman, who with one long smell left her madly in love. He had a lazy gaze and his body was tight and fibrous as the sweet heart of a sugarcane. His reddish skin reminded her of the tops of the mahogany tables in the Velazquez house. When Eladio Salaman drew close to Marina that night, he arrived with a tidal wave of new fragrances that left her enraptured for hours, while he led her by the arm all around the town square.

The ground of the rain forest, mint leaves sprinkled with dew, a brand-new washbasin, morning ocean spray ... Marina began to practice the most difficult odors to see if she could invoke Eladio Salaman's. This effort drew her attention away from all her other duties, and at times she inadvertently served her masters dishes that had the wrong fragrances. One afternoon the shrimp and cassava came out smelling like pork chops with vegetables. Another day, the rice with pigeon peas perfumed the air with the aroma of greens and salt cod. The crisis reached such extremes that a potato casserole came out of the oven smelling exactly like the Velazquez boy's underpants. They had to call a doctor, for everyone in the house who ate that day vomited until they coughed up nothing but bile. They believed they had suffered severe food poisoning.

Marina realized that the only way to control her fascination with that man was to see him again. Secretly she searched for him on all the town's corners, using her sense of smell, until two days later she found him sitting in front of the Serceda Theater drinking a soda. That afternoon, Marina invented an excuse, and did not return to the house in time to prepare lunch. Later she ran home in time to cook dinner, which was the most flavorful meal that was ever eaten in the Velazquez dining room throughout the whole history of the town, for it smelled of love and of Eladio Salaman's sweet body.

One afternoon, while strolling through the neighborhood, Hipolito saw the two of them, Marina and Eladio, hand in hand, smiling and entwined in each other's aromas. He remembered how the dark woman had rejected him and now he found her lost in the caresses of that black sugarcane cutter. He waited for the appropriate moment and went to speak with his esteemed mother. Who knows what Hipolito told her—but when Marina arrived back at the house, Doña Georgina was furious.

"Indecent, evil, stinking black woman."

And Mama Edovina was forced to intervene to convince the mistress not to throw her daughter out of her house. Doña Georgina agreed, but only on the condition that Marina take a cut in her wages and an increase in her supervision. Marina couldn't go to the market unaccompanied, she couldn't stroll on the town square during the week, and she could only communicate with Eladio through messages.

Those were terrible days. Marina couldn't sleep; she couldn't work. Her vast memory of smells disappeared in one fell swoop. The food she prepared came out insipid—all of it smelled like an empty chest of drawers. This caused Doña Georgina to redouble her insults. "Conniving little thing, Jezebel, polecat."

One afternoon Marina decided she wouldn't take any more. She decided to summon Eladio through her scent, one that she had made in a measured and defined way and shown to him one day of kisses in the untilled back lots of the sugar plantation. "This is my fragrance," Marina had told him. "Remember it well." And Eladio, fascinated, drank it in so completely that Marina's fragrance would be absorbed into his skin like a tattoo.

Marina studied the direction of the wind carefully. She opened the windows of the mansion and prepared to perfume the whole town with herself. Immediately the stray dogs began to howl and the citizens rushed hurriedly through the streets, for they thought they were producing that smell of frightened bromeliads and burning saliva. Two blocks down the street, Eladio, who was talking to some friends, recognized the aroma; he excused himself and ran to see Marina. But as they kissed, the Velazquez boy broke in on them and, insulting him all the way, threw Eladio out of the house.

As soon as the door was closed, Hipolito proposed to Marina that if she let him suck her little titties he would maintain their secret and not say anything to his mother. "You can keep your job and escape Mama's insults, too," he told her, approaching her.

Marina became so infuriated that she couldn't control her body. From all of her pores wafted a scaly odor mixed with the stench of burned oil and acid used for cleaning engines. The odor was so intense that Hipolito had to lean on the living room's big colonial sofa with the medallions, overwhelmed by a wave of dizziness. He felt as if they had pulled the floor out from under him, and he fell squarely down on the freshly mopped tiles.

Marina sketched a victorious smile. With a firm tread she strode into Doña Georgina's bedroom. She filled the room with an aroma of desperate melancholy (she had drawn it from her father's body) that trampled the sheets and dressers. She was going to kill that old woman with pure frustration. Calmly she went to her room, bundled her things together, and gazed around the mansion. That pest, the Velazquez boy, lay on the floor in a state from which he would never fully recover. The master bedroom smelled of stale dreams that accelerated the palpitations of the heart. The whole house gave off disconnected, nonsensical aromas, so that nobody in town ever wanted to visit the Velazquez house again.

Marina smiled. Now she would go see Eladio. She would go resuscitate the Pinchimoja. She would leave that house forever. But before exiting through the front door a few filthy words—which surprised even her—escaped from her mouth. Walking down the balcony stairs, she was heard to say with determination,

"Let them say *now* that blacks stink!"

Brine Mirror

Exactly alike, deliriously so, in all our parts. The motel, the smell of motel soap, of gas from the air conditioner, the smell of disinfectant and Spanish cider and her, naked against my skin my stomach cramp, my nakedness which never seemed so identical, so repeated, my exact cramps in her stomach, talking, talking, talking. We know she talks when she's nervous. We know that we can't shut up because then the tongue, the kisses, exactly alike. We know we have to shut up, but I don't know when.

The bodies we share are men, once together, often separately. We search among the skin that's not the same, it's nothing but the same. Those skins don't have alleyways of pores that we like to have. They don't have nipples standing straight up with radiant aureoles expanded—the little warts on top looking up, black like food. They don't have areas of skin asleep on their forearms, like ours, nor calluses from writing so much. They don't have soft hands and hard hips, but the reverse. They don't have a soft almond hidden away, oyster nut with its possession, but the reverse. About that possession, I say it and I finally look her in the face, we together and I am reflected on her tummy which also rumbles. Out of fear we are alone tangled in the smell of a motel room. We look straight at them, those two, who look directly at each other this time.

Exactly equal, we shut up to better confront lips, tongues, hungers.

"And then what ... then what do we do ... then why this fear?"

"It's time for us to shut up," I say and I put my tongue between lips like two oysters, two creatures absolutely of brine.

It's the day before and we are talking on the phone.

"Today he didn't want to look at me. He was fixing a painting," I say.

"And he didn't even wonder if it was me who was kissing him?"

"The other guy loves me too much, as if the world were a satellite revolving around my ankle. I don't want them to adore me, I only want ..."

"Exactly."

"You see?"

Then she explained that she had told him how I felt and I thought it was or wasn't me she was describing but herself, who got another call just then and I realized while waiting on the line that we sometimes spoke with someone else the same way, while she waits, with me there.

"What should we do?"
"Let's go drink at Pipe's Bar."
"Not today, he's coming by to see me."
"Okay, well, then I'll go to the movies with the other guy. Did you see *The Piano?*"
"And what about us?"
"Well, tomorrow."

It was a week earlier and she was arriving on an airplane. I was waiting for her. I wanted to see her with the scarf I bought her before she left, I wanted to see her dressed for the cold, to see if ... To see if that would hide her hips, the huge ass just like mine, the fat legs, the little waist covered by currents of dark skin going down to see. I wanted her right there, fresh from flight, hair uncombed, just reconstructed in memory and down the hall of the Caribbean International Airport. Looking precise and composed, with her Roman Color #37 on her mouth her tight jeans with my scarf on her neck, as I would come to her, as I have arrived to her after flying. She disembarked from the airbus, arrived in the baggage claim area. I waved from the other side of the glass. She put her mouth against the glass. I kissed her on the mouth in front of everybody, through the glass, without realizing it.

We are two clams, I say, and now I can't stop the choking and all I have left to do is moan between her flesh or is it mine, her hard nipples in my mouth and I don't know who moans hers so small me afraid of hurting her so hard her flesh so hard her hard thighs her big buttocks and on top of that I don't know if I'm darker and bigger everywhere that we are inside her mouth and I put my tongue where she puts her salty tongue in the bitter salty mouth like me and tasting. The cramps of fear in my lips that go down my legs, hers, my hands I don't want to see it confuses me to open my eyes I want with my mouth, motel, motel of a dammed sea and hairy fleshy inside an excess of rivers with a grotto and a sucking I touch I hear moans one finger small ones and long ones two three almost the whole hand I can only see the back of it there all the hands fit her little fingers have to wriggle and we break she repeats turns me gets on top of her long hair it smells like a motel it smells like a motel like a mirror it smells so black sea motel and a mirror with brine, it's her hair that confuses her face which is not the same however but and it's her nipple my her mouth her Roman Color everywhere which is the same one I use, our faces come undone inside hers come undone her inside me undone inside undone.

It's the day before on the phone and my mouth already going down. It takes her a whole day to arrive. The hands warmer, fresh asphalt.

"What do we do now?"

We come back from seeing Aurora and I take you to your grandmother's house. They're going to put her in a home, you tell me. I don't want to look at you, hands and heart sweaty. They're going to put her in a home you don't get out of the car I can't wait for you to get out so I can escape 90 miles an hour on the expressway to get home and shower and watch television without looking, I don't want to look. You don't get out of the car. They're going to put her in a home and you look at me, you look at me and I have no other choice but to look at you and reveal my fear of admitting that I'm dying to put my tongue there where those words are coming out, all the words, the ones that burn.

"They're going to put her in a home, but you're talking about something else."

"Thinking about it too."

"What do we do now?" I turn on the car, sighing.

"Let's get out of here."

The Parks

She knows, she knows what she's going for
She knows, knows

She's lost the way to her grandmother's house, to her house, she's lost the way to get to the school and the way to the mall, and the way to her friend Lucy's house. Poor girl, she's lost them all. But the not-getting-there is so delicious. She pretends as if she's going, oh yes, she pretends that she's in a hurry even, or that she's hurrying in a distracted way—so sinuous. She is the perfect picture of a young girl who walks astray, fifteen years old and going through a forest, a forest in the middle of the city; how convenient, a forest like a mall, like a pool in which one learns synchronized swimming; how convenient, a forest like a park, like a movie theater when she says: "I'm going to the mall." And she goes to the park. Oh, poor girl. "Be careful in the parks," say the voices of parental guidance. And she's smiling, she sees the clouds smiling at her, inviting her for a walk.

She has a plan in her head and she's fifteen years old, and she has some Pepe blue jeans that the last old man that she met in the park gave her. She stands on a bench under a lamp post, takes out a Maja brand powder compact under a lamp post, arches her back and pretends to look at herself in the mirror. She has a projection right between her eyes.

"Mister, can you give me a cigarette? They don't let me smoke in my house." And the man gives her a cigarette, gives her the fang of a white dragon, gives her a caterpillar full of milk that explodes among her hair while she settles on the rug of the park, the green rug among the plants. She settles and she gets down on all fours and she stretches with the old man inside her—he's wearing a condom of course, she's not crazy, and she hears him moan and she moans, the old man giving her gifts while she gives herself to him and gives herself to others.

Oh yes, to others. Because she does not go to the park to get gifts, nor for the old men escaping from television and their wives to give her milk caterpillars and then clean her up worriedly with papers as green as the park. She knows that other creatures live in the parks and that these are invisible, creatures of pure projection.

The crackling of some nearby plants, the motion of shadows far from the lamp post, directly far away, but close, the precise distance away to be able to see how she opens herself and the old man thinks it's for him, and it's not for him. There are some eyes without mirrors that look at them. More than one pair, seven pairs, one thousand eight hundred pairs, a whole sea of eyes that cover the road that borders the park. She imagines those eyes that are on motorcycles speeding towards supermarkets; those eyes driving towards night jobs, stop for one second at the park and on her little arched back, and the old man that gives her caterpillars, get relief for one moment, a very small moment, and then they continue on their daily routine. When she thinks about those eyes, pleasure becomes a delicious storm of chimes on the grass. Oh her, and her very same body is the salad and the main course for those eyes. Oh the old man, who is the spice, and it could be any man if taken in the appropriate doses.

Sometimes, walking aimlessly, she runs into friends from the park. She has seen them in the shadows, they have seen her looking and letting herself be seen for an instant. Later they return to each other the secret of their zippers. They don't touch her, those don't touch her or say anything to her. She smiles at them. They calm her down. They don't have names, their name is not Tony, for example, who used to be her boyfriend and who never understood why, even when they had enough money to go to a motel, she would want to go to the park. "We can get mugged here. Any neighbor of my dad's could see us here."

And he wouldn't let her, he wouldn't let her really open up for the others who moved around in the shadows stalking her breath, shaking her little arched back, licking her with the tongues of their eyes. He wouldn't let her get chewed up inside and eat herself to become a spectacle, to show herself, to see herself, to spit it out, to speculate a little scream, a moan from inside; Mr. Tony was an accessory of the parks, the park and its eyes on her. She is fifteen years old and she gets lost in the woods like a mall, she is the star of the mall, with carefully cared-for leaves, trimmed, milky, oh the milk from the park, with its eyelashes they bathe her from afar. There are twenty pairs of balls, there are twenty pairs of eyes who spy at her and who turn her on all fours; they despair at what they see, what the shadows reveal, they despair when they see the girl's skin doubled up beneath a lamppost and with little park scratches. Tony was a nice guy, but he should move, move out of the light because there are eyes there, there are eyes there. "I think they're looking at us."

And she's going to blow up inside, all of her inside. The rest of the night is an effect of all the parks and beaches and roads where there are eyes; she has seen them everywhere, there are eyes without a face, jerking off in the name of that little piece of meat that shines whipped and sweaty on the corners. There are eyes and eyes that come against a dark sky and who fill the stars with milk and they wax the sidewalks with their milk; one gets splashed with milk at seven on the way to school, there is still fresh milk hanging from the lamp posts, shining on the asphalt and the benches of the bus stops. The old man in the car at the stoplight with sunglasses and his hand going crazy, right under the steering wheel, letting her go by. The neighbor behind the patio of her own house, looking at her through the blinds and through the leaves, leaving her gifts. Four glasses of milk a day for strong bones and shiny hair. She's fifteen years old and blooming.

Tony got tired of her little game, so she found other men. Her friends at school did the park part-time to buy designer jeans and satin blouses for parties. At home she had enough for that. But she didn't have excuses for the park. "Mom, I'm going to Lucy's house." She already had a few fixed old men, and it wasn't with just anybody, and they weren't bad at all. They gave her enough for excuses and for whatever she wanted. They gave her little gifts and her park in the late afternoon, almost at night. They protected her from whoever tried to go too far, violate the rules of her game. One of them was a security guard.

She is fifteen years old, has a park close at hand, has accomplices, and once in a while she goes for a walk. She gets good grades in school. Someday she'll go to the university. She likes Math and English. She doesn't have any vices, she's not stupid, and in the afternoon she closes her eyes for one instant to see herself as the image that the eyes peering into the park see, with her little back arched and her Maja brand powder compact like a Delilah looking out ... to start the game. She goes out on the streets towards the park, she pretends to walk there, but they better not wait for her. Deliciously she loses herself through the little bromeliad paths, through the gardens of Maltese cross. The eyes always find her to leave her their gifts.

Stained Glass Fish

She walked in out of mere curiosity. She had always wanted to go, check out the ambiance, see if what they said about it was true. She paid the cover charge, approached the bar and ordered a vodka and tonic. She glanced around. It was a club like any other, dimly lit, with high-tech decoration—lots of aluminum, the bar a steel color, smooth as a jet of electricity. In the back of the place, over towards the dance floor, there were quite a few couples, drinking, smoking, screaming in each others ears over the danceable techno-pop music. Some of the various couples kissed hastily; they ran their hands obliquely along waists and backs; they danced too close, "just like any bar," she thought, calming herself down.

Then she saw her. "Damn, it can't be!" she murmured quietly. She thought of what she would say if the other woman saw her there: *I came out of mere curiosity. I've never been to one of these places.* Maybe she would greet her as if there was nothing to it, as though they were in a normal, common place. She turned her back to the dance floor, she faced the bar and began to observe her in the mirror. She was still sitting, leaning against a stained glass window of a fish made of reds and blues that seemed to swallow all of the light in the club. She was beautiful, with a long oval face, just as a centaur's fingernails must look, and a loose mane of curling hair, like an immense pubis. Yes, it was definitely her, her colleague from work and even from a few power lunches. She was smoking. She was dressed in black, with leather pumps and a low-cut top. Her firm olive-skinned bosom, her breasts which must smell of hidden things, of glass fish, of salty ocean spray, she thought, of shipwrecks. She was talking with another woman; she smoked and laughed deliciously.

Juliana looked at her slowly. *Out of curiosity,* she explained to herself. She stopped to look at the solid legs, the solid bust, her presence, that woman's solid laugh. She rested and recovered enough to order another vodka and tonic. The woman disappeared from her field of vision for just an instant, which was occupied by a dark hole from which overflowed a compact, lipstick, Kleenex, keys and finally the wallet more or less pregnant with photos and—*there's no reason to feel guilty*—money. She paid and returned to her original position to continue observing. She watched as she shattered the ambiance with a dagger's intention. She leaned over towards the other woman, whispered something in her ear and kissed her softly on the lips. Then they left together, holding hands, to dash themselves against the night.

Secretly Juliana felt all the hate in the world bursting up through her legs. Confused, acting macho, she knocked back the rest of the drink that remained in her glass. She immediately rose from her stool and left to get her car from the parking lot next door.

She got home somewhere around 1:30 in the morning. An apartment full of furniture, in pastel colors, with lots of paintings deliberately hanging out of order on the walls; some antiquities on the bookshelves; mirrors. She went straight to the bathroom to take off her makeup. She covered her face with a white cream that smelled just like the ocean spray that filtered in through the balcony on windy days. She turned on the water and let it run while she looked in the mirror. Juliana thought of her, or more precisely, of the moment when she smashed the bar's ambiance in two irreconcilable pieces, took the other woman by the hand and exited, leaving the stained glass fish confused by such an excess of certainty. She replayed the scene over and over again, the image remaining identical each time, until she was exhausted.

Juliana rinsed her face and turned off the water. She went to her room and searched for a nightgown. She undressed. Accidentally she rested her hands on her breasts for a few seconds, firm and dark breasts just like the other woman's, breasts that also smelled of something remote which could never be adequately named. Slowly she forgot about the nightgown and she fell into bed naked. She touched her neck, her pelvis, all the secret places that only a woman recognizes. She drifted back into her memory. This would be the precise instant in which *she* would return to her house after making love, to clean off her makeup, to undress, to fall into bed and touch those same hidden places Juliana was touching. Then she understood that the hate she had felt rising through her legs in the bar was envy.

Juliana suffered all that night, and the following day and

night, trying to convince herself that all of this was normal, that her loneliness must be to blame for this deviance. She believed she had rested from the tempest a few moments before sunrise, sure of her reconciliation with her regular patterns of behavior. Determined to go to work, she waited for morning, dressed, gulped down a cup of black coffee and marched out to her car in the parking lot.

She arrived at work at 7:45 A.M., took the elevator to the eleventh floor, and walked to the office. She entered. There she was, beautiful, dressed in black, her mane of hair hanging loose, her face with the look of a sudden sigh. She was smoking, and it seemed like she was still smoking the same cigarette she had been smoking a few nights ago in the bar. Juliana walked quickly by her, to avoid her gaze. But she couldn't. She was offering Juliana a broad smile.

Almost running, Juliana shut herself in her office. She looked for forms, files, she copied papers. She finished her backlog of work—and not even two hours had passed. Then she began to suffer again, just like the night before and the day before and the night before. She searched for relief, and convinced herself that her salvation would be in thinking only of trivial details—how much water she should pour on the magnolias that afternoon, what she needed to buy at the market, the dress she wanted to put on layaway in that elegant store. She thought of going to some party, to some restaurant, wearing that dress, of inviting some colleague from work to the event. She thought of her fellow workers, such idiots. If she invited one of the men out for a drink, he'd surely misinterpret it. He'd either think of her as easy or as interested in special favors at work. She thought of the dinner she'd cook that night for herself, alone.

Arriving at this point, Juliana couldn't help but think of her. She thought that maybe *she* would understand intuitively the true dimensions of Juliana's anguish, of her incomplete ferocity, of her solitude. She felt there would never be a chance of being understood with anybody but her, for *she* surely knew how to make that precise fragrance gush forth, that personal odor, natural and remote, that smell of a stained glass fish that inhabited the breasts of all the women in the universe.

As noon approached, Juliana firmly opened the door. She walked towards the work room. There she was, the other woman, studying some plans. Juliana walked over to her, she tapped her shoulder and waited until her mythical face appeared among the curly shadows of her hair. Then, determined, she opened her mouth to ask:

"Would you like to have lunch with me today?"

Dilcia M.

for all the Puerto Rican political prisoners jailed in federal prisons by the U.S. government since 1980

All day locked in her cell, she had forgotten where the day began. The food tastes like walls, and the walls like a vagina and her vagina was another one of those walls where they had implanted a video camera to monitor her slightest breath. Her desire to walk was monitored, and her desire to touch any orifice and her sleeping and her eating and everything. She doesn't remember anymore when this started; she has ceased counting. Now she can't even imagine where her name began.

They caught her like they caught all of them ...

Above her desk (she has one now) is a picture of her from senior year. In her senior year—Chicago High School, Central—she strutted around in Ditto jeans and a striped tee shirt and boots and a down jacket during the winter. In the photo she glances out with a lost gaze, through that soft focus they always add to senior year photographs, and she's wearing too much eye shadow. In the photo (*Dilcia Marina, 1973*) she's beautiful. She remembers her boyfriend and how after the senior prom they rode in his car to the beach near the river. By the river sometimes the wind would lift her skirt, and sometimes her boyfriend would lift her skirt. There was a warm breeze that night. The river silt was like the real sand of any other beach, like the beaches that they told her were real, the beaches on an island, on the island that was sometimes hers. Her boyfriend lifted her skirt higher and she let him. She let him caress her, she massaged him at the same time on the sly, and the boyfriend kissed her neck, her breasts, her belly button which he made into a beach. And to her, lying there on the beach of her own belly, it didn't matter if the guards watched her twisting under the sheet while she remembered all this—the sheet was a river, an artificial beach. It didn't matter to her if they increased the volume on the camera to hear her groaning and if they jacked off in her name. It didn't matter to her. She put her fingers in where occasionally the guards searched her, and she said, "This is mine and it doesn't matter if in a few minutes they come to take it away from me; for now it's still mine, mine."

She finished quickly, afraid they'd come and interrupt her like they sometimes did just to embarrass her. With her hands still wet she picked up the picture; with her face still wet she looked at it. She looked at it, and with fingertips still wet she traced the face in the photograph. She knew that she made love to herself, to the Dilcia Marina that she had been when she was seventeen and who she was now beginning to be, again, in that cell. She lay down to sleep. She dreamed of many bewitched things, of that day when they came to arrest her. She resisted, but they caught her. Scared, she got up saying that word that at times loses its meaning—the word for which she set herself to learn to fire and to construct a false floor and to store and transport the stolen M-16's, .38's and .45's. Inside that room she had definitively forgotten where that word started, where its roots are and where it ends.

"You have a visitor." Dilcia M. stood up, surprised that she could walk. Do her feet begin on the floor (the false one)?

"The conditions in prison are almost unbearable, but we will continue struggling for ..." She stopped writing the bulletin. She didn't have the spirit anymore to keep playing with the words she'd memorized for these occasions. Although now she really didn't know when to use them—when the psychiatrists and reporters came to interview her (all monitored on video, of course), when she answered the letters that people wrote her, or when they strip-searched her after visits—those words which she grasped with the last claws of her conscience. She knew that now an enormous nostalgia would overwhelm her. It happened every time that she began to write those words, those words for which they had captured her, like all the others.

She got up from the desk and walked around the cell. She couldn't help but ask herself how the people who saw her photo in the leaflets would imagine her—that is, in the copy of a photo that was taken when she was free, senior year of high school, when they still used the flip look and she spent hours in the bathroom with a blow dryer trying to look like a Latina version of Farrah Fawcett. Once the group sent her a Christmas leaflet which quoted her alongside her photo; she was that same woman who was so brave, who said all those things without the smallest doubt and swore she was going straight to her pantheon before she would renounce the cause, who continued calling the enemy "enemy" and assuring the people (who?) that she was fine and that she was paying for the consequences of her sacrificial act, which is in itself an act of love. She couldn't believe she was that courageous woman, the one who spouted all those convincing words. And she asked herself how many people would read that pamphlet, how many would she convince, how many people would she have to carry on her conscience. "Dilcia, honey," she shouted, "you are such an idiot!"

Some time had passed, two or three hours, or maybe it was dark out by now or the sun was rising; she didn't have access to a window this time. But some time had passed. She knew that. "Out there ..." Today, for some reason, she was thinking about the world outside. Today she couldn't stop thinking about the world. Surely there was someone that was sad, preoccupied with her and her manner of surviving. But who? She had heard rumors that a few comrades had begun to propose negotiations so they could be freed on their own recognizance. The oldest resisted the most; they had resisted for the last twelve years. She didn't know if ... It was better to think of the world outside. If they can bring enough pressure to bear, maybe; if the party, if all the parties, if the Civil Rights Commission, then ... maybe; or if something would happen to her and she didn't wake up one morning, then she would have fulfilled the sacrifice. The years are too many, the struggle is too arduous, and she isn't so heroic even though she memorized (in the last scraps of her memory that remain) how to pretend to be brave almost automatically. One act of love, another act of love, but why shouldn't she want something back? Outside there is someone who maybe understands that she needs to get out of that cell right now, before she turns into a traitor, because now she is even forgetting how to pretend.

Act of Faith

Contents of the purse confiscated from Ms. Blanca Canales on October 31, 1950.

A black leather pocketbook, closed by a zipper on its upper face, a simple style, with medium-sized handles.

It contains:

- a social worker's license;
- two checks donated to the Puerto Rican Nationalist Party;
- forty-five bullets, .38 caliber; and
- a postcard of Joan of Arc which on its back is inscribed, "Saint Joan of Arc, intercede for Puerto Rican independence."

Oso Blanco

for Awilda Sterling

Wake up, go to the bathroom, wash your face, your mouth, have breakfast, brush your hair, get dressed in a hurry, go out, go out, go out, turn on, key, where the fuck is the … turn on the car, garage, take it out of the garage, close the gate, be late, always late, be late because of the damned traffic jam. Drive with anxiety, run into a never-ending line of cars face to tail; and there in the distance, the jail. From there it's five minutes to the office. There it is. Five more minutes of torture and that's it, it's over. That white mass, high walls, almost falling onto the avenue, in the middle of the highway, almost. Her watch. Her girlfriend from work, over there in Caguas, whose son was put in jail once, drugs, there he was, she went crazy, almost crazy; she would forget everything halfway through saying it, her memories would die mid-tongue and she didn't know what to say, she didn't have anything else to say. He was there, inside one of those little boxes so he wouldn't hurt anyone else. A mass in the middle of the expressway, look, a white mass with its protective wires made of blades, and its wall, little pigeons. Five more minutes.

Arrive, in a hurry, in a hurry, parking—there isn't any, same old shit, go around the block to try behind the pharmacy, get out—the key, don't leave it inside like last week, idiot, the key and the purse, put on lipstick, brush hair again, let's see, the mirror, my face is already shiny, I look like a shining pot, I look like the hood of a car at noon, where the fuck is the powder compact ... more lipstick, more things that hang from my arms, the keys, get out of the car. Do I have enough deodorant on? Get out of the car, put on "The Club," the car alarm, punch the clock, punch in on time, only seven minutes late, record time. Work ... work ... and more work ... work, stamp papers, write reports, answer the phones, coffee break, go down to the cafeteria, coffee, a sandwich, a diet soda, I'll take the newspaper, up to the office, the elevator overflowing with messengers, with secretaries, with hands and arms with bags made of brown paper, memo paper, letter paper, payroll paper, one hour for lunch, fill out applications, move inventory, listen to the radio, digest the lunch special, make the time go by. Look out the window that opens over the expressway, there, far away, almost erased in the distance, the white walls, little birds.

Four-thirty, five, get in the car, once again pass by the jail. That mass there, immovable, tall like a seagull right in the middle of the expressway—not like a seagull, no, like an elephant—no, not like an elephant, like a heavy white bear, that's it, a polar bear, *oso blanco* standing on its two hind legs like the circus bears, the circus clown bears full of Purina and vitamins for their eyes, so that they don't fall off their

tricycles, from the high wire, a clown bear standing on the wire of the expressway ... right in the middle. It's already tomorrow. It's already tomorrow, and tomorrow
(brush hair, get dressed, look for keys)
the jail waking up over the expressway
five more minutes and the morning traffic jam will be finished and tomorrow
it's already tomorrow
the day after tomorrow
the day after the day after tomorrow it's already tomorrow
(a bear doing somersaults on the expressway)
the day after the day after the day after tomorrow
a light
hand peeks out in between the bars of a cell
and starts to wave. Again, again, another. (the car goes by, the hand comes out at exactly the same time, not before, not after) another (is it for me?)
another (is it for me?) another (It's for me.) Get up, brush your hair, get fixed up in a hurry, the keys, damn the ... ah, I left them right here, the keys, and the deodorant and the jewelry that goes on the left hand. Today I'm going to buy a new ring, a ring that reflects the sun, and the things for the car, take it out of the garage, lock up, run, fly to the expressway, get into the traffic jam, wait, pass in front of the jail, wait, pass in front of the clown bear jail. Wait, wait in the traffic jam, the cell wakes up, the arm comes out, the arm comes out and recognizes the mountain green car, old and falling apart, it recognizes the green, old, falling-apart car and smiles, she had never seen an arm smile, but now the whole arm comes out and smiles and waves every morning

(maybe it's her friend's son). She dresses her hand up once a day. She buys rings and bracelets that shine in the sun, she paints her nails, she lifts weights to tone her arm for the arm that waves from the fourth cell from the left, the one that faces the expressway.

She doesn't know why that arm is in jail, nor what other marks it might have, nor what other movements aside from the slow gravitating from side to side that caresses her, her, her, that will caress her through the air. Maybe that arm has killed, maybe it has strangled innocent young women necking with their boyfriends on roads far away from the noise and the light and the eyes of the city, maybe it has grabbed knives, guns, maybe it smells like gunpowder. But it looks so far away and so harmless and so hungry for affection and so graceful in the sun and so dark and muscular and strong and sweet and chatty and unknown. No, no, it can't be the arm of her friend's son, she would recognize it, this is another arm, without memory, without a past, that is born there every morning, for her alone, for her alone (she remembers another arm that once ...) The arm, oh ... that arm doing inexplicable things to her, from the car, green and falling apart, the arm that smiles at her and caresses her from the air. And like that, nothing more, the arm flies, flies from its cell, climbs onto her arm and begins to undress her of all her little rings, of all her sorceries so the sun can shine on the skin, her skin the skin of the mountain green old car that's falling apart, the skin of the mountain of anxiety that she carries held tight inside, in the callus on her elbow, the arm flies, the arm grabs her arm with its hand, with its fingers, scratches her with its fingernails and wets her with the sweat of so much waving in the air. The arm rubs her arm. It lets itself slide and fall onto her lap, the lap gets hot, the broken-down lap, mountain green, the lap and her mountain underneath, beating, ah, after so long, the lap and the longed-for arm, she waves, she doesn't want the people next to her to see. It's hers alone, that prison arm in her lap, folding up the pleats of cloth, making the cloth of her pleats swell up with heat, a heat like the one you feel when you awaken alive again with an arm on your lap, oh, the lap and the arm that already touches her skin lightly with the tips of its fingers, her soft skin, her skin that once knew about these things. Once knew. But that was a long time ago, a long time ago, she suspects it was a long time ago. She was little, smaller than when girls start to know about those things. Yes, or maybe she dreamt it once as a girl, that there was an arm that made her feel like now, that would teach her to spell things on paper, to spell things on the open pleats finger by finger, open mouth, lips, tickling and grateful tremors, through that arm, her skin learned things that when she grew up she denied knowing; when she grew up she stayed alone and absurd because everything else that she began to feel could never compare with the arm, now alive, under the pressure of the fingers, he has her, in the prison of fingers under the marks of hard nails. She keeps waving so that they don't notice anything, so that the people behind her won't honk their horns in the traffic jam, so that nobody knows about the slight pressure of her foot on the accelerator, the sharpness of her other barefoot heel on the brake. Her heel swells up with blood and the arm swells up in her lap. There go the fingers looking for another pleat in her skin, under the underwear. The fingers touch the wetness. She sweats, and she doesn't know if it's the arm that's also sweating, the arm that escaped, she helps it in its escape, the fugitive arm, hidden in between her legs, puts its fingers inside her, touches her firmly, another finger, and another, wriggles around, wanting to split her open with joy and it doesn't know, it doesn't know of the happiness that it's drawing on her, her arm almost makes her crash into the family mini-van in front of her. She stops in time, but then lets herself go, lets herself go. Fingers go inside her, lips swell up and so wet she holds in a little scream on the tip of her tongue, the finger's prison is the moss between her legs, and her clitoris hard like a seed, sending shocks through the whole system, each hair on her skin stands on end, she lets herself go, she recovers, and again, a vertigo in her head, her nipples harden, she opens her mouth, arches her back, lets herself go, the arm takes her on flights, drives her, takes off her brakes. She leans her head back, rests it in the small headrest on the drivers seat and goes off in contractions, in bubbles of humidity, she floats through the air from the fingers that fill her with tickling sensations, grab strong on the steering wheel, get close in the air. The other fingers of her arm inside there take her prisoner, flesh against flesh, pleat against nail, against sweat, inside her mountain green, broken-down car.

    smoke
    parking
    turn the key
    go in
    close the garage again
    again ...

Her green car inside, and inside the house, the room, the bed, to dream again with the beloved arm. It's already tomorrow.

II.

This is my plan. Little by little I have been convincing my cells to separate imperceptibly from the cells of the shoulder. It was hard in the beginning, because first I had to convince the brain that all of this, really, was *its* idea. Clearly, scientists are mistaken. Not all reflective activity takes place irreducibly in the brain. There are other parts of the body that, given the correct circumstances, can in fact perform these operations. The legs, for example, don't only respond to neural stimuli that come from above; rather, if separated, they can act on their own. I imagine that being locked up, and having lost the tertiary faculties of the brain, caused the strange condition that I have mentioned to take place. I imagine that is what happened to me.

I have consumed many hours figuring out this mystery. I still do not have a sure answer, but one thing is undeniable. All of a sudden, that is to say, on a day like any other, my fingers started to form consciousness about themselves, from the tips to the nails to the cartilage of the epidermis. Each digit came to life independently, phalanges, carpals and metacarpals. And it wasn't that they became hyper-sensitive to the stimuli that are always present, it wasn't that the faculty of touch evolved in such a way that it could pick up more sensations than before. It was that they had taken on self-awareness. They didn't even need to feel anything; they could imagine it in the abstract, break it down, and articulate it in a non-linguistic form, rather an electrical form (to put some kind of name to the nervous stimuli) that could sustain intelligent and intelligible conversations with the rest of the conglomerate that comprises me. They could remember, invent concepts, analyze.

In the beginning, each part—elbows, skin, hair, cells, fingers, triceps, biceps, ligaments—would send each other messages, among themselves and separately. It became the Babel to end all Babels. Imagine each cell, each commissure of the skin and bones talking at the same time. It took some work, but little by little we established order and set rules to all emissions. Then we read and understood each other perfectly, we were in harmony, discovering ourselves as living beings, capable of reflection and dialogue. A marvel. Unanimously they chose me, both singular and multiple, as a regulating entity. Since I form and am formed by each one of those parts, I possess a broader consciousness of how we are connected to each other and to the rest of the body.

I should also say that the rest of the body did not know what was happening, nor did the brain, nor the eyes, insensible as they were in this confinement that progressively subtracted our faculties from the total entity, sectioning off each one of the parts. It was strange, because the more conscious I became of myself, and the more I became aware of my identity, the more the rest of the body was losing itself in a profound lethargy.

In the beginning I took measures to awaken my companions. I began by sending electrical signals to the internal organs, but I soon noticed that these did not go beyond the shoulder. A strange border had erected itself there, one that did not permit communication. I continued sending messages upon messages—tongue, are you there? answer me, thighs; nose, nose, can you smell me? At the end of the month the only member that answered me, with a weak yet lucid signal, was the brain. It was not difficult to make it my ally. I gave myself the simple task of earning its trust, responding once in a while to the silly signals it sent me—scratch a calf, hold a glass and take it to the mouth, hold a pill under the tongue and with the fingers extract it and throw it far away, kill insects, take the fingers and squeeze the penis (which would swell up) and massage it up and down vigorously until it overflowed, spitting out semen. We never felt as alive, the rest of the body and myself, as during that milky moment. The buttocks would tighten up with fury, the back arched on its own, a current of spikes went through the whole skin, every hair would stand on end, and the mouth opened, mute for years, contorted from pleasure until it, too, expelled, a long groan that came out from the center of us all. But those were reflexes, not thought. There was no introspection—only gluttony; no abstraction—only sensations. Every time this happened, it reinforced my certitude that the only one who thought about that conglomerate of flesh, sweat and hair, was me.

Even the caresses turned automatic and boring. I sunk in a bottomless solitude that seemed worse than all the daily tortures of the guards, the frequent vexations from the other prisoners every time they took us out to the corner to feel the sun. So much solitude, so much self-consciousness, for what? We had to turn to the outside. Sometimes electric waves would reach me that translated the sounds that the ears meekly picked up. "This one is pretending so he'll be sent to the psychiatric hospital." I thought perhaps that maybe *that* was the solution to my problem: that we all be moved from that filthy cell to a place in which other body parts would be going through the same thing, a location that could serve as a meeting place for arms, legs, eyes, mouths, or livers, that would be the body's last stronghold where something similar to thought would be taking place. Who knows if in that place there would be other organs where that operation, which previously took place in the brain, would have taken shelter. That is how I began to develop my theory, which I dreamt about sharing with someone out there.

There were days of doubt. Maybe we are the only ones who have been through something similar, we who live and form this imprisoned body, this obviously sick criminal body. Maybe we are, maybe I am, a mutation, the first one, undocumented and therefore lost in the abyss of scientific oblivion. I swear that during those days I would have given anything to have eyes. That way I would have been able to cry openly about all the desolation that invaded me. Since I cannot, I order each follicle to sweat almost to dehydration. I want to dry myself, dry myself forever and not fear anymore, not feel anymore, not be aware of anything at all.

It was one of those days that I saw it. Well, *see* it as such, no; rather, I felt it, that other arm leaning against a metal hole that was stuck in front of the jail. At first I could not believe what my skin, my fingers, the whole surface was feeling. It was like a morning heat, a flight, a warmth that overtook my whole being. I tricked the brain into ordering the total to approach the barred window; and later, desperate to know, to connect to that living thing, I stretched out from the cell and made myself wave, signaling that brother that was saving me from the existential abyss that consumed me. It took three days to successfully make the connection. But finally the other arm answered. We stayed that way for a long time, caressing each other through the air, until it disappeared in the distance.

This event repeated itself frequently and filled me with joy, with passion, with I don't know what other feelings coming from places that were unsuspected by me until that instant. Each day that went by, the other arm would decorate itself with accessories. I could perceive certain bands of a different temperature around its fingers, around its wrist; a chemical reading revealed pungent substances that covered the patina of the nails. It was a very cute arm, which waved, and which was obviously adorning itself for me.

I overflowed with joy. I could not believe the miracle life was offering me. The palm of my hand was sweating waiting for the other hand, the one that waved daily in the distance. Reacting to my happiness, the rest of the body also came up with responses. For example, one morning I saw that the mouth was showing its teeth in an easy and peaceful way and that it was curving the lips towards the sides, trying to reach the ears. It was smiling.

One night when the penis was swollen with blood, I surprised the brain, retrieving that arm from its visual memory: my love, my beloved. That vision waved and waved in the air. The skin got goose bumps, the body temperature went up, stored images of fingers over foreign vulvas surfaced, hidden sensations in the lower abdomen. My hands and my fingers started remembering things. In that very short moment of brain conspiracy, every cell in the skin recovered, each one in its own way, the memory of the textures of other bodies' skins, unusual encounters with humid membranes, viscous solutions that tangled fingertip to fingertip and slid down wetting the whole hand; remembering unusual temperatures, tremors and swelling on his long-ago digits going into pleats of skin that I would have never imagined existed. When the brain wanted me to go to the pubis and begin to massage, a colossal fury overtook me and I started hitting the face, attacking it with electrical messages to make it faint. I didn't know what was happening to me. On the one hand I wanted to keep feeling that banquet of sensations that made every follicle stand on end. On the other hand, I was dying in anguish that another organ degraded me by using memories without consulting me, leaving me with the horrible alternative function of feeling everything from afar. Crazed, I rammed the poor face, the eyes, the cheeks. The lungs were full of air; the mouth, the throat started to scream. I knew it wasn't those organs' fault, instead it was the fault of the damned brain, but how to get to it without hurting the rest of the body? The mouth started to scream. I couldn't control myself. Because of the noise, the guards came and with blows quieted all of us down. I got several blows, which contributed to my peacefulness, more so when the brain tried to protect itself with me, calling on me in bits of messages, so that I would cover it from the blows and the kicks.

After that incident I didn't communicate for days. I didn't want to know anything about that filthy traitor folded up in between the bones of the head. And "that one" holed up to me as well and refused to send signals to the rest of the body to direct us to get close to the window. Every morning, as hard as I tried to cross the enormous distance, I couldn't do it, not even the day when a commotion formed out there (I felt the sound waves on my skin) because of something that did not want to move from in front of the cell, an artifact inhabited by another body with the left hand extended out towards the prison, that waved and waved awaiting a signal.

III.

I'm the vicious bear that devours bodies from the prison. I'm the star of the circus. I turn into windows, I turn into bars, sometimes I pull out guts, and I'm a quite a magical bear, a trapeze-artist bear, a juggling bear and a jail and a penitentiary. I have never been able to eat anyone from the outside. But there's always a first time.

There was a succulent arm and another one outside who waved at it and I tried to eat both of them. Ah, you might say, what a bad bear, what a sadistic, cruel, fetishistic bear. As though *you* weren't, you who's reading the memories of the trick performed in the insides of the nasty bear. Polar bear, artist of the expressway—oh, as if your mouth didn't water thinking about flesh and hearing about tight buttocks and sweaty skin, and the aches of rubbing and entrances and exits and desires that never get fully consummated. You are vicious bears. You, too, do tricks and eat bodies and gorge on milk and honey. I know it, I know it.

One day I ate two prisoners who were playing doggy next to the boilers, barking, one mounted on the other. They were naked like dogs, although dogs don't go around naked because they always have hair; these didn't have so much hair, that's why I say they were naked. They played doggy, barking and mounting each other. I ate them, for their indecency. Humans cannot play like doggies, humans should not get naked next to the boilers, touch their flanks, fill up with blood, squash their lips together and ferociously bite their jaws. Humans should not wet buttocks with saliva, put the tip of their dicks through that little hole, they should not be pleased at seeing it expand, only to end up stuffed with so much meat. They should not be enveloped in that aroma of something that is almost rotten, but alive, baby shrimp stranded on seaweed. They shouldn't. Well, thinking it over, other humans can do that, but not prisoners. The prisoners are my food.

I was looking inside myself that day and I saw them. My way of digesting is unique. I am a clever bear, I am a circus bear and the star of the show. That means that anything that has to do with me is unique. And what a magic trick I operated. If you could see them, my little dogs, one encrusted in the buttocks of the other, one pulling out blood and shit and the other one roaring, who knows if from pain, pleasure or fury. He was small and the other one was forcing him. Every time he got sent to the boilers, he shook in fear. That's where the other one waited for him, he knew it. There and in any dark corner, wherever he pleased, when he felt like it. The little dog bit, squirmed, screamed, but he always ended up under the grip of the bigger one.

In reality, it wasn't just those two, there were more. A pack, a whole multitude of prisoners playing doggy, with the little doggy not letting himself be taken. When the big one finished, another climbed on top of him. He asked his friends to hold his hands and legs. The little dog had fallen on the floor and the others tore what flesh was left on him to pieces. They opened his legs, his mouth, they held his arms and one by one they shoved their dicks into that little hole, already blood red, in pain. The little dog roared and cried, cried and roared. I ate them whole, I called the guards, who, disgusted, starting beating all of them, including the little dog, and crushed them for my hungry teeth.

But my favorite delicacy was in that remote cell. It was a beautiful jet-black prisoner. His offense against society was unique. I felt such brotherhood with him. When he was free he would take girls and stick his long fingers up through the little doors that were just forming on their childish parts. What a magic trick! What circus hunger! When he was free, he had been a school teacher, he had taught those little girls to read and write. He held the soft little hands in his big hands, firmly and expertly, hands with pencils willing to confront the traces of their names, spelling out the identities of the people that were going to mold their little lives. In his huge hands, the little hands battled with wood, with graphite, with paper and the rest of the vegetable substances, willing to name things with signs. Their tender hands became more childish each day, they found it harder each day to separate from the little elbows, the skin, the little arms dirty with thorns and sweat. The big hands started playing with the little hands, then large hands with little thighs, and finally the huge hands started unbuttoning pants and little blouses to play with other more tender flesh. Flesh that didn't know the surface of a pencil, nor even the rubbing against paper. It only recognized the cloth that covered it and now the teacher's huge hands, that thought they were little hands and went into those tiny holes, "don't tell your daddy that we play this game," that opened holes for each one of his fingers, "your friends don't have to find out." The huge hands were careful not to hurt them, nor to leave marks in the tender, little, desired holes. It was enough to just smell, feel the heat and the humid skin. That's why it was so hard to trap them. Years went by, before some of the little girls were grown up and recognized what the huge hands of the teacher had shown them, those perverse and tender huge hands.

My delicacy was in maximum security, so that the rest of the prisoners wouldn't try to eat him. That's how I arranged it, the clever bear. He was there by order of the warden and he never or almost never went out. Once or twice a month, the teacher bronzed in the sun in a corner of the patio and the prisoners dragged him to a hidden corner to play doggy with him. But since he was lost in his memories, he offered no resistance. He suffered everything with the temperament of a martyr. And I thought, how beautiful my teacher, my delicacy, how lovely; he's a saint, an angel, an innocent devil, a big child who doesn't know what happened to him, why they punish him, how ungrateful love is, what pleasure and what a delicacy. I saved him for later and I started devouring him slowly.

But then it was that naughty arm, that suicidal arm, that arm, gymnast of the air. With its accomplice outside, it interrupted my digestion. AGGGHHH! I roared exquisitely. I am a mighty bear, I am the circus star, I am the king of the expressway and I plan to eat you. No naughty arm is going to take you away, not even bit by bit, my succulent delicacy. No sneaky arm is going to take away fingers and nails and skin memories that transport arms to legs without panties in the back room of a school. AGGGHHH!!! I roared exquisitely.

Luckily I am powerful and voracious, luckily I'm a lucky bear. I did a mean trick and I laughed HA HA HAAA!!! ... in the middle of the expressway. The gymnast arm fell out of favor when I turned the rest of the prisoner against it. Its ally from the outside stayed waiting for it in the middle of the expressway, causing the greatest traffic jam in all the world's history of traffic jams in this part of the Caribbean. You might say, *Holy Virgin! What a confused and cowardly bear, what a monstrous and voracious bear*. As though you weren't enjoying my show, eating my words, imagining my little dog prisoners and gymnast arms and sweat. As if you didn't hide behind your big eyeglasses to try and devour me. Admit it, you are just like me. You are just like me.

Nobody can beat the nasty bear, the bear of the snowy fields, that lifts palaces on its hind legs and balances on the very slender cloth of the expressway. I am invincible. I am the best. You can't beat me. I am a very clever bear, and I defend my food well.

Mystic Rose

for Harry Hernández

Like ink, your hands taste like ink, and your whole neck like ink, the whole thing, like ink, and he laughed as he always did and moaned deeply while she sucked his fingers one by one, one by one she pulled the ink out of his nails, with her tongue, with her teeth. He thought he would get dizzy. He came covered in ink from the workshop; yesterday he had made another artist's proof, adding red to the wooden slab. He wanted to give it to her, to get close to her that way. But she wasn't interested in how the paper's fibers swallowed the colors, but rather in the ink in his own fibers, on his hands, his neck, his hair. And while she would draw into her mouth all of his stained skin, he swore that he was getting lost in little colored lights that would suddenly appear before his eyes.

Colors. He found them one by one, putting them on the printing slab ... He made more incisions on the wood, each line deeper, and he enjoyed it, how he enjoyed encrusting the metal into the surface, taking out more threads of cortex

... more ... more ... He would put colors in there that she provoked with her tongue. Time flew by. Another sheet of paper. I'm coming over to show you the changes that I made on the self-portrait. *Come directly, don't go by your house, don't change, come directly here.* And again her tongue, the proof lying on the sofa and him, wriggling on the living room floor under her tongue, her tongue, her tongue taking the color out of his body.

With gouges and wood, that later turned into stone, plastic, acid on metal, needles, skin, that's how he began tattooing his body. First on the inside of the wrist, then on the left nipple. She waited for each tattoo with wet lips, she would jump on him, drag him to where the colors were, mountain green ink, and red. The black crust of the loose skin in each incision, falling on her tongue and dissolving. The needles gilded his body, spitting colors into the skin. It hurt, it hurt intensely, so densely that the piece of flesh became numb, and with her tongue she would make it hurt more, moving him to delirium that she later relieved by scraping the fresh tattoo; the itch after each bite. She would suck the fresh ink from his skin, she would swallow him by parts—and then he would leave. She wouldn't call for weeks, didn't answer messages. Until the next tattoo.

Winged serpent, tribal dolphin of the Inuits of the northwestern United States, Celtic band on the ankle, star of David, the sun of the Incas, Year 5 of the flower of the Mayan calendar, three waves, two dots and a bar on the back side of the neck, Egyptian scarab, Taíno fertility symbol, her tongue licking, her mouth sucking—the third eye and its pyramid, salamander of a medieval bestiary, an Abakua death mask, scorpion on the ear, teeth on the ear nibbling, blood falling from between the skin, blood and ink, *no body can resist this*, four Japanese ideograms, with her on top and the moaning. Moans that fill him with colors. I'm coming to your house, I have a new tattoo, but this time you have to find it. She waited for him in the apartment. She undressed him slowly and looked for the fresh ink. She couldn't find it. There, the known inscriptions, the colors that during the last months she knew by smell, touch and sight. The little bumps on the skin that raised under her fingers. She looked and looked as he smiled, satisfied.

She began to lick the tattoos in order, one by one—*you taste like ink, you taste and smell of ink, my food, where is my new food, tell me tell me*—and him, still and smiling, like a squid stewing in its own ink, in dizziness with bubbles of color. She continued in order of their appearance, the inside of the wrist, the nipple, the ankle, the back, the calf, the left forearm, the back part of the neck, *no body can resist this, no body can resist this*, he swore he was dying and she kept on going down, her tongue kept on going down, her fingers searching his skin, her tongue went in between his legs, he smiled and didn't say a word, he wriggled and sighed and fought with the air in his throat and with the tangled sheets. She turned him around by force, searching everywhere, *tell me, I can't wait, tell me tell me*—she turned him on his back again. He could see everything black, greenish black, drowned; with her hands she began to separate his buttocks. He stiffened up, his body stiffened up involuntarily, it was burning. Very expertly, she waited for him to relax, and went towards the same place slower, this time he let himself be opened, this time he let her lick him, lick the pink rose of the seas that they had tattooed on his pinkest flesh—the most virgin of all, that flesh that he knew bore likeness to that other pink flesh of that other mouth with which she would draw colors out from his marrow, from the bones on out, everything greenish black and blue and red. *Here, here it is.* She would eat him and he held in the screams in his chest to not scare her, he was biting the pillow, biting himself and felt the beloved taste of blood and ink on the tip of his mouth. He held open his buttocks himself, he separated them for his beloved's tongue, so that she could really see where the needle had gone in, so that she could check the strength of the trace, the looseness of the line, the volume of that pink rose of the seas, so that she could get shipwrecked with him inside there, inside his colors. She was eating him, eating him and swallowing him. He was going to say *I can't stand it, no body can resist this*—but when he opened his mouth, he felt a sea of ink that submerged him, he had no time to say anything, he didn't even have time to believe that he was dying, because in fact, he had died.

That same afternoon, she went with the tattooed corpse to hand herself into the authorities. She waited for the results of the autopsy, knowing that she had killed him by sucking the blood from his soul. She told them in detail how the game with the skin kept growing and about the almanac of the mysterious inscriptions on the body of her dead lover. She cried seas of mascara, declaring herself at the end of the confession: guilty. The results of the autopsy—septicemia. They let her go, referring her to a psychiatrist and without filing charges. Crazy. Who could believe such a story?

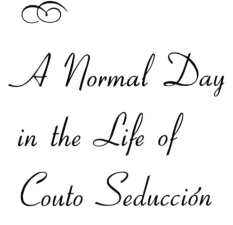

A Normal Day in the Life of Couto Seducción

What you see here is his bed, and these are his sheets, still warm from his sweat, and this is the place where his shaven head lies, his beautiful head, perfectly round, framed by two ears, each wearing an earring of the finest gold. Only the purest things may come into contact with the body of Couto Seducción.

Each morning when he wakes up, Couto Seducción has his bath prepared for him. It is his custom: his beauty ritual begins in the big, antique, porcelain bathtub with lion's paws. Polishing his skin, which is very white with a map of filigree veins that paint his geography blue, in this way he gives himself to the sun, to the bees and to the air. He is all generosity. His body, it is enough to glance at his body to know it without a doubt.

And his beauty is millennial. People comment that Couto

Seducción comes from an old Portuguese family that emigrated, thanks to the work and grace of chance, to the town of Manatí in the Caribbean, a town full of salt, the same salt that swelled the flesh of the first Coutos and tutored them in the art of seduction by mere presence. For that is the tradition of this lineage. Making himself present before fine spirits, who also adored the maddest pleasures, was enough cause for desire to be freed in torrents. A Couto arrived at any gathering, with his slow pachyderm walk, as if chosen by the gods—legs like two cornucopias, huge belly and bellybutton and wide waist where whole cities hung—he arrived with the face of absolute satisfaction. That is to say that, when any of the Coutos showed up at any gathering, it looked like time stood still to see him go by. The earth stopped its rotations, even hummingbirds stopped batting their wings so as not to miss that flesh that invited the saliva of every living being, object or mineral. He went as he was, clean, smelling of all the fragrances of the world, with eyes that looked like fresh almonds (a genetic characteristic of the Couto Seducción family), and pink lips. One could swear that those lips had just eaten of the forbidden fruit, that one was in the presence of the Chosen One.

After the bath, Couto Seducción dries off his skin with freshly washed towels that have been hung to dry in the night wind. Afterward, still naked, he opens the doors and windows of his room and lets the sun lick the whole continent of his skin. He carefully chooses his clothes for the day. Nothing ostentatious—a brightly colored pair of pants, a cotton shirt—so that he, not his clothing, is the *seducción*. In that way he rigorously follows the lessons and secrets that his ancestors bequeathed him, a perfect work of art, the culmination of generations of professional seducers. He looks in the mirror, approving his outfit, and walks to the kitchen. He is going to have breakfast.

It matters very little what Couto Seducción eats each morning, as it matters very little what he eats for lunch or dinner. Because, contrary to what any heedless observer might think, the unquestionable size of Seducción is not due to what he ingests in food. Whoever sees him without looking, would confuse his amplitude with fat. But not us; myself, who completes the group of his lovers, we know that Couto Seducción's broad body is pure enjoyment, pure generosity.

We are thirteen lovers. And all thirteen of us happily wait for him every second Tuesday of the month, here in his family's old mansion. We come from different parts of the world, all seduced by the same thing. The last one to join us was a Swedish woman who, after seeing him strolling on the public beach where she was camping on vacation, left everything, everything, to give herself the satisfaction of having him near, of loving him with her little body, and her little legs and her tiny, common mouth. She joined us knowing that if there were thirteen of us (there is no number more cabalistic), the wind, the salt, the sand and the rocks would conspire to give the chosen one pleasure. His body is so vast—so perfect and so vast.

It is because I am one of his lovers that I know all this I'm telling you, and can describe his bed and his porcelain bathtub and his ritual. I can tell you more, I can describe in detail how Couto Seducción spends the rest of his day: doing what everyone else does. Working as a researcher in a pharmaceutical laboratory, and even chatting with his co-workers, some of whom have formed or now form part of the secret company of the thirteen who love him. I can describe to you how he returns to his house and lights a cigar and smokes it, looking at the sunset, and how the second Tuesday of every month he lies about some doctor's appointment (and the boss believes him, everybody believes anything from Couto Seducción, however absurd it may sound) and they let him leave by mid-afternoon. He then gets into his jeep and drives toward the old family house, where no one lives. There we are, all of us, waiting for him, on time, full of desire, ready to achieve what until now we have never achieved: to completely satiate him, to exhaust him from pleasure. His body is so vast, so perfect.

Sometimes he surprises us with some detail. One day he arrived with seven dozen yellow roses that he wanted to strip of their petals to make the perfect bed for lust. Another Tuesday he came carrying the most delicious *almojábanas* that any living creature on the face of the earth has ever tasted. We played a perverse game with the *almojábanas*. We dipped them in honey and tried to totally cover him with those delicacies. Needless to say, it was a failed endeavour. He is excess, the perfection of excess, and that became clear that month. For March, he wanted us to make love—the thirteen of us and him, for whom there is no number—on the seashore. That is how we lost one of the lovers, who, determined to kiss one of his buttocks, rolled under the inmense mass, and there lost air and consciousness. But we can give testimony that he died happily: the current took him with a huge erection that looked like the fin of a marauding and carnivorous fish. Once we were able to reconstruct the confederacy (we succeeded at the end of June; we are very selective about who can share with us the exquisite body of Couto Seducción), our lover dressed up as a woman and let us lick her breasts and let the males among us penetrate him through an enormous women's sex that (who knows how) he managed to make with the pleats of his pubis. And today, naked, all of us, hungry, all of us, we await our lover, asking ourselves what new surprise he will bring in addition to his insatiable and infinite body.

It is two in the afternoon, and that one who is approaching, walking with the tranquillity of a divine being untouched by time, is Couto Seducción. He is already climbing the stairs one by one, he's putting the key in the lock. We all smile. He arrives with a cobalt-blue glass bottle.

"I brought almond oil. Today we will play that we are slippery like fish. Who wants to start?"

The Writer

Monday, 3:35 P.M.

"*He approached the mirror and didn't see anybody. He tried washing out his eyes; surely some fuzziness was fooling him. He stretched himself out over the sink. He found himself a little to the left, or the right; the image still moved away from him.*" The writer reads out loud, she revises word order and accepts it. She likes the way 'fuzziness was fooling him' sounded with a pause, with a brief pause that leaves space for the other, for the bit about the guy leaning over the sink confused like a devil at Angelus. That's how she had seen them, her men. And—well, it's not that she has seen them that way, but that she remembers, or really she desires, or better yet that's how she remembers wanting to see men, splitting themselves in half before a complication. Looking at the guy you can see the confusion and the sweat and the dizziness of his lips, and even better if he has a mustache, so that he himself understands that it doesn't work as camouflage. Yeah, the guy will have a mustache. "*Bit by bit the image focuses in his eyes, helped by the hand which travels from the wash basin to his face, to the cheek moistened by cold sweat, the head so naked of emotions that it amplifies the noise of his hand rubbing unshaved stubble, pausing on the mustache that hasn't served to shelter him from the terror for which he hadn't premeditated a word, nor a little grin nor a cute existential saying."*

Maybe it would be best to end the description here. Maybe, because it could be that a longer description would serve to give form, dimension, to the character. But then how can she justify the narrator's voice that knows what's happening to this guy as though it were an omnipotent guru? The voice of the scribe, the scrivener, the scribbler now converted into Writer-Visionary, center of knowledge, and by knowing so much, close to God, to Thatcher, if she writes. Which to choose?

(Which do I, writing this story about a woman writing about a guy rubbing his mustache in front of the bathroom sink, choose? I, the liar, the one who already knows the ending [bullshit; the one who dreams of an ending which as yet escapes her]? Will I make an effort to give the impression of having everything under control? Or maybe I'll lean towards the trick that I don't have *any* control, that this story writes itself alone, as if the computer had gone off on a journey, its circuits overloaded, all-powerful, very close to God, or Thatcher, if it writes.

And here again I am feigning. I am also assuming a voice that is not the real one, but the one prescribed for the character of a standard writer already expert in the craft of storytelling, playing at being confused. Me, the liar, trying to choose where to put the commas and the double-entendres—

and as far as the descriptions, how much life to give them, how many personal and imaginary experiences, and how many other writers to include as models. [And I don't name them so as not to implicate them in this voyage toward words.]

And now, without fuss, I ask myself what goes on inside a story, how to mix this uncertainty with the story of the writer who tells about him and her, the guy and the girl, Marina and Abnel, the lover about to break up, the story on the edge of not being, the voice of a woman speaking about another and even another, all of them beginners in this game of words and all of them nevertheless extremely magical. How am I going to tell this story, from what authority? Which things will I keep up my sleeve and which will I write in this circuit through which I travel dressed as three people, yet naked? The story will make it fit.)

4:15 P.M.

"Mommy?" Listen to them call her. The kids are back from school again; her peace has ended. "I'm here in the study." If only they would arrive a little later, if someone would take them out for a ride, if her husband (now ex), who was so quick to disseminate, would give her a hand. But oh well ... from the hands that he gave her, tarnished with who knows what kind of stains, grubby from having fingered so many other surfaces. "The car has to be waxed ... Baby, I can't ... I come home all tired out from a hard day's work, and now you come and ask me if I can ... Honey, what you mean I should take them out? It's Friday and the boys are waiting for me ... Me, after I've given you everything, you *dare* to raise your ... ?"

Hand? Yes, the one that he trained by force, the one that made it impossible for her not to protest, but that at times slipped out of its harness. Gentle hand of ironing and transcribing recipes in the kitchen. Hand made meek from resting on the impatient back of a husband in heat. Gentle hand for combing babies' hair. Mild hand of offices and paperwork. Hand that gently, bit by bit, wanted to evict the other one, the rebellious hand that touched herself between the legs and grabbed her shifting pelvis, that wanted to throw the long locks behind her in an attitude of defiance, the hand that wrote "fugitive ideas" in the corners of application forms and work applications. That hand was the one who finally threw an immaculate "get the fuck out of here" gesture, without scratches or blood or burns, with the fingernails perfect—and the one that she, so hopeful, regarded with the eyes of a terminal patient, confident that it, the rebellious hand, would save her from total collapse, would pull her from that idiotic lethargy which was sucking her life from her.

The writer hears her children calling her, asking for something to eat. Food? It's making itself, and they better wait; she's not a flame. If she were a flame ... Today it's better for them to leave her alone. And after a whole afternoon of working on the damn story, now the territory to be mapped is even greater. More work, as though she didn't have to earn her living, go to that private office, make thousands of sacrifices to finish her master's degree in the evenings, and return to the house to cook, iron, mop, and take care of the kids. As if she could shut herself up in her room as that English woman advised her, that would be easy, and set herself to writing.

"Marina, what's up? Why are you resisting me so?" she asks the protagonist in her story, remembering how she wrote in her student years. Good stories. She had shown them to her professors. In those days she hung around disheveled; she wrote things with an astounding facility. She had even been published once or twice in literary magazines and everyone predicted a clear and brilliant future.

(I imagine the writer young, hungry, very good at screaming in a loud voice and on paper the things that bother her without a bit of shame. I imagine her readers, her colleagues at the university, friends, one or two young professors reading her story with a bit of irritation at first, but later becoming more interested, encountering errors, but overlooking them now because they are so involved in the tricks of the anecdotes and the precipitous use of words jammed together on the paper, undressing each other and camouflaging each other all at once. That's how it is when you are young, with the hunger of wanting to stuff everything down your maw and swallow it all without even chewing it. "It needs editing," they would have told her, as though she didn't know that. Will this story need editing? Will I turn out to be very young to be writing it? How many years must pass for one to have accumulated enough schizophrenia to be able to stand in front of the text and start using the scissors without mercy? And I doubt the functions of editing, especially in that part of the story about the writer and her life, about her battles with time, her children, her work, her ex-husband.)

"God damn the moment I met him. God dammit," she grumbled, and she remembered her ex a decade ago, when she saw him appear with his financial face, his mustache that tickled even the most disinterested back, his amorous expertise, swelling that curriculum vitae between her legs, enchanted and without any suspicion of pregnancy. "I'm going to get your lunch right now. No television until you finish your homework. What's that aw-Mommy, aw-Mommy—Start studying!" she shouted at the kids, while she thought about that pothole on the road of life that was the guy she had just finished divorcing.

She still liked him. Even now if she ran into him on a corner, she'd fall all over him like a rainstorm. Although the truth was that these days she felt so dry. Ever since the divorce was official she hadn't slept with anybody, with the exception of that guy from the office who invited her to dance and who she opened up to with such fury afterwards out of hunger and the surprise of seeing herself in bed with a man who wasn't *her* man (excuse me, her ex-man). That body so different, that situation in which she discovered herself of not knowing where to put her legs or on which side to lie—what most surprised her was that while her colleague from work jumped up and down on her, she found herself thinking of her children, of the empty refrigerator, of the check due on the fifteenth, of the story. It made her want to cry, and the guy from the office thought she was super excited, reaching the sparkling moment, so he accelerated and melted inside her, confessing his pleasure and pushing, pushing ...

(I, the other, find myself pushing and asking myself why I don't dare to edit this information, the portrait of a life so different from my own. I, who have never been divorced, who doesn't know what it's like to be pregnant, nor how this fact connects with the story of Marina and Abnel, the lovers. Thinking about not knowing what giving birth is—although from the time I became conscious they have been explaining it to me over and over again. Thinking about not knowing what writing a story is, although I have examined the x-rays of so many of these wordy creatures, from Poe discussing Hawthorne to Cortazar's *Breves Apuntes*. Someone once told me that writing is like giving birth. Some help *that* is!)

5:00 P.M.
"*He turned towards the wall. He thought of Marina naked, sitting on the bed, slowly smoking her Marlboro Light. He thought of how he had followed the long outline of her body in saliva, the pleasure of a cartographer addicted to delineating the borders of any surface. He thought (oh yes, he was sure) that she didn't even suspect his intentions, those that were unveiled from the moment he saw her. He, Abnel Nieves, draftsman, would leave her soon, alone as she could be. Since then he had been planning his rejection ceremony right up to the day when he would gather up his last shirt and say goodbye to her. Even before the empty mirror he remembered her with his fingertips; he passed them across his stomach and down to his pubis. He began to transform himself into his character, the lover. He smiled with great sadness, just as she liked the protagonists in her favorite stories to smile, and he walked upstairs to see if he could collect the outline again.*

"*Then Marina told him, 'I'm pregnant, Abnel.' She told him, 'Maybe now that you got a job in that firm of agronomists ... I mean this wasn't in my plans.' She went on telling him, 'Me, just when I'm about to graduate, one more semester and I finish my bachelor's degree—but you know that with a bachelor's degree in*

Literature you can't earn anything. That's why I wanted to go to graduate school, get out of this country that suffocates everyone, that leaves you abandoned in the street. Marriage? And when my old man finds out—! But no, I'm an adult. What can he do to me? You tell me—what? Hit me? He's threatened me so many times, it wouldn't be any big deal if he finally did it. They threw Mom out of the house when she was sixteen ...'

"Silence. It didn't occur to him to fill it with anything, not even with a caress. He's like he has never been before, scared to death that a woman will ask him for something he can't give: encouragement, company, his true identity as a scared and hurt man before this turn of fortune. But who would think of getting pregnant now? Maybe she hadn't taken all of the necessary precautions? Now he would have no choice but to hate her, hate her with pity, for a mistake. Marina continued speaking, 'Abnel, I'm pregnant ... it'll break Mom's heart in two if she finds out because she always wanted another chance to live and I have been giving it to her. Nononono, I can't do that to her, Abnel. Maybe the best thing would be to tear the little thing out, if it's so small that it doesn't know it exists. It won't matter to him; it's for Grandmother, who since she was born hasn't had a life, dammit. Abnel, you understand me, right? If you put up half the money, I'll go tomorrow morning ...'

"He suffered from nausea and sudden chills. He had to get up and go to the bathroom while she talked and talked of her plans and her mother, of abortion, of how much they had told her it hurts, that it was like a vacuum cleaner that they put in between your legs and it sucked out everything. She talked about the complete cost, about his savings, and he wanted to vomit, tear the little baby out of his stomach, tear Marina from his body, tear out the six and a half months that they had been navigating the mid-afternoons in the bed in his apartment, tear out his sense of touch, the marine flavor on his tongue, vomit all over her and wash his saliva off her forever. And on top of it all, that abortion that she contemplated, talking to herself. Alone. Marina had left him alone. Abnel noticed that she could care less if he was listening, if he loved her, if he understood her. She had arrived there that afternoon prepared for the worst, completely distrusting that man who had gotten her pregnant; she came to demand payment for her afternoons, for the shock, half of the money for the abortion and then, ciao, nice meeting you, see you around, and a farewell kiss.

"Marina talking, talktalktalktalktalk, 'and well, you know, you don't even have to take me because Nandita offered to go with me. The thing is, I'd feel better if it were another woman, if she grabs my hand while they connect the vacuum cleaner and I say good-bye to the little creature. I don't even want to imagine it; I don't want to know how its little mouth will be, and its little hands, because don't think that I don't dream about it, that I don't have feelings, it's just that ...' She was nervous, captivated by the texture of her own voice, exorcizing the spirit of the baby, saying good-bye to it from this moment. And his presence, that of Abnel Nieves Contreras, was unnecessary. He could have been any other man standing there, with his back to her, leaning on the bathroom sink. Holding on as if he were on the edge of an abyss, and with his lungs racing. 'And I'm going to kill it, I know. That's not what worries me.' He couldn't stand any more, but if he screamed, she would look over and notice him. She'd discover his initial plan, that what seduced him about her in the first place was the possibility of abandoning her. Then she would dress slowly and leave him, leave him. No way; it's the cartographer who always leaves.

"Nevertheless, that abortion in the mapped territory definitively closed his exit. His dreams of being a macho but not abusive man vanished into thin air. After working so hard to be a free man, not a scared little boy looking for his place, pleasing authorities and papers, insuring his employment, respect and mobility. After working so hard to abandon himself to temporary pleasure and rest from the demands of that courage that he never exactly had located, nor from where it sprang when they summoned it, precisely now ... He had to improvise. If he denied anything, even if he never saw her again, Marina would continue to be sitting there naked on his bed, smoking, talking to him about a little tiny baby that didn't even know it existed and about a mother, old and sad, thinking of her daughter with the hope of living through her.

"Abnel leaned closer to the mirror and saw no one. He tried to clean his eyes; surely some fuzziness was fooling him. He stretched himself out over the sink. He found himself a little to the left, or the right; the image still moved away from him. Bit by bit the image focuses in his eyes, helped by the hand which travels from the washbasin to his face, to the cheek moistened by cold sweat, the head so naked of emotions that it amplifies the noise of his hand rubbing unshaved stubble, pausing on the mustache that hasn't served to shelter him from the terror for which he hadn't premeditated a word, nor a little grin nor a cute existential saying. He felt a burning rising from his belly. That burning suffocated him violently. He recognized that it was hate, hate for poor Marina, hate for the unborn baby, a wrongful hate that came to torture him."

6:30 P.M.

"Sara honey, turn off the beans, they're about to burn. It's the button on the front, the one on the right." Let the beans burn, let her life burn, the story was burning up her life and she had to finish it. My God, what an obsession she'd let get under her skin.

(To me, who always has defended other voices, those of the metaphor. Why this sudden urgency to narrate? And what do I know about how the writer feels, what part of her mind she uses to open herself up to unknown eyes? And what am I doing inventing this disguise to discuss the problem of writing if you're a divorced mother with children, or a single woman with a lover, or a plain woman, or noncorporeal as I pretend to present myself on this page, through this game of distance? How much do I pay for undressing myself before the eyes of the unacquainted, letting them see bumps and cellulite and those immense buttocks that my voice possesses, and the hunger to want to swallow this story which I've been working on since who knows when? Let this end now, let this end before the words dry up on me.)

The long nights in a monastic habit for the damn story, she, who for years did nothing but imagine the incredible pleasure of sleeping alone again in a wide and uninhabited bed; of walking naked around the bedroom without eyes that said to her 'turn around, baby ... that's it ... let me devour you with my eyes;' without hands that tried to touch the most distant layer of her epidermis, without the fear that some husband would open the door to search for matching ties or unmatched socks or little bottles of aftershave. NOBODY IN HER ROOM—this was the longed-for definition, the one that persuaded her to divorce. Afterwards this story grew, sentence by sentence, and in the beginning that made her happy. But then, the bedroom filled up yet again with eyes, with hands, almost without her noticing. She had never been so cornered, not even when she knew she was pregnant, not even when she found herself getting married to a stranger, not even when she gave birth to that little bloody creature and then another, not even when she saw her husband going by in the car with other women, not even when she knew she was divorced and she found herself suddenly in that room with a life so incredibly new that she started it crying from terror and happiness.

Now, the story filled her with frustration, with anguish for having betrayed him, *him*, the versions dreamed between the snores of the financial husband, the versions discarded as cockroach food. Because it was she who betrayed them, who betrayed the stories she carried inside. She left them alone and without paper or walls to write themselves on because she was busy chasing after a man she didn't even know, by not wanting to plot, by being afraid to shut her mouth and listen to her own silence. For years she complained about him, years of blaming him for her failure, for her desperation, years of *hailmaryfullofgrace*'s, *youareanabuser*'s, *youaintworthnothing*'s, *ifonlysomebodyhadwarnedme*'s—years. Hating him by mistake. Years, because it never occurred to her (imagine that!) that she had the right to assume her own guilt and continue forward, without asking for any more forgiveness than the necessary, without demanding useless compensations, without wrapping herself in the role of a martyred mother and a denied wife.

7:45 P.M.

The writer served dinner to the kids, checked their homework, turned the TV on for them. On her way back to the study she considers how things would be if she could talk to her, to the girlfriend of the guy with the mustache, to Marina, about what her story is really about. And she would like to talk to her—like a friend, or like a potential friend, like a friend turned into an enemy, or like a friendly friend. She should sit down in front of her one day, some rainy Sunday in the Bombonera to have a coffee, and a mysterious dialogue would unwind itself between them, where they would talk about how much the price of coffee had gone up and about vegetables you could eat to lose weight and about how much they like fancy underwear and about something that both of them know has happened to them, that which brought them together in the dark alley of the story.

WRITER: Look, Marina, I don't want you to be my reflection, nor my creation, nor the prolongation of my life, nor that you be another. I only want to know what you are good for and why I have persisted with you for so many years. I've done everything to flee you, but it's useless.

MARINA: It wasn't my intention to cause you so much anxiety. And actually, I don't even really know what you're talking about.

WRITER: About what your function is.

MARINA: Does there have to be only one?

WRITER: Yes.

MARINA: But *your* life has more than one purpose? Well then?

WRITER: But you're not alive, not like me. Oh, let's abandon this theme, because we'll end up talking about levels of reality and all that philosophical garbage.
MARINA: Whatever you say; you're the live one.
WRITER: You feed off of me, right?
MARINA: Right.
WRITER: Why me? Why do *I* have to be the source of your nutrition?
MARINA: But, what a determination to think of yourself as indispensable! You're necessary but insufficient, like a plumber, like a sewer, or like a can of soda out in the fields. I feed off of you, you off of me, but in addition we feed in another way. If what you want is the certainty of what you're doing, that this story is the final purpose of your existence, do me a favor ... go talk to another character.

8:00 P.M.

"Marina approached him from behind, while he continued to grasp the bathroom sink like a drowning man. 'What do you think? Can you put up half the money? There's a clinic here in the area where they don't charge more than $300. Now you know, don't worry about going with me because Nandita said she'd take me. Anyways, I'm going to stay in her apartment to recover.' Don't resist her at all, let her decide. All of this made him want to cry; what he always wanted was a little love affair with fixed commitments, planned, in which the ending would arrive slowly and painlessly.

"Crushed, conscious in his mind; in silence he walks over to his pants and pulls out the checkbook. 'Make it out to Cash or in my name. *Tomorrow I'll go to the bank and change it so I can have the money in my hands and pay in cash.'* He, Abnel, almost with tears in his eyes, makes out a check for three hundred dollars. She looks at it. *'It's okay. You're studying and I work. I'll pay for the whole thing.'* He wanted to joke. *'You get the next one.'* But he couldn't even laugh at his own joke. Marina walked over to where she had her clothes piled one on top of the other like a mountain. While he watched her put the check away in her purse and begin to put on each article of clothing with incredible precision, Abnel asks himself what good all that practice did, all that preoccupation with acquiring a sad smile from a story.

"It's true that he was able to get Marina to open her legs happily without worrying about anything, converted into a mere surface. But now she was escaping from him without any remedy. Marina. She observes him in his loss. Marina. And that baby, even if she aborts it, will remain in some corner of her head; it will demand nightmares, odors, strange stories. It will give her another dimension, it will make her deep, it will fill her with things that he cannot and does not want to include in his cartographies. Her eyes with be populated with questions and everything will be lost, the almost seven months of arduous labor to leave her empty for him and his abandonment, gentle and superficial, uselessly wasted. Abnel knows this is the farewell; he knows that never again will she let herself undress for him, not in the way he wants her to, never again in the way that he needs to see her get undressed so he can assume his character as the lover."

MIDNIGHT

"Something is going wrong," the writer thinks. Wasn't the story going to be about Marina? Then why does she keep feeling compelled to describe Abnel's thoughts? Now it's gone on for pages and pages and still ... Why can't she encounter the transition to the girl, who's still faraway, smoking, naked, her Marlboro Light glowing in the shadows of the story? The writer asks if she'll have to start all over again, convinced that she is following a logic that can't provide the guidelines to understand Marina. The writer prepares herself to develop a list of guilt, an inventory of reasons that explain the reason why a solution escapes her. Maybe then she can understand her block.

LIST OF REASONS
1. *the nine years without writing;*
2. *the kids who don't let me ever complete my chain of thought;*
3. *the study where I write is very small and stuffy;*
4. *work;*
5. *the bills;*
6. *it's the divorce and the ex's fault, for being a week and a half late with the check for the kids;*
7. *it's the ex's new boss's fault; and*
8. *it's Marina and Abnel's fault, for not letting themselves be written the way I want, especially Marina, who doesn't notice Abnel's plans: to turn her into jelly so he can play his games of saliva and routes on a map. She's always thinking of nothing but her Mom and the fucking abortion. In others, always in others, thinking of the others. Let her go on and finally get rid of the baby and stop annoying so much. After all, it's no big deal. I would have done it if it wasn't for Abnel, who didn't let me, who invented the whole bit about marriage to confuse me. Yeah, Marina's the guilty one.*

The writer stopped writing and reread her list. She felt a punch of blood in her chest, a pressure, an intense desolation. She understood that there was no point in continuing to work on the story, at least for today. She stood up from the chair. She walked towards the kitchen where dinner's dirty plates awaited her. But before leaving the study, she returned to the typewriter, she turned it on, and she added one final item to the list:

9. *I am guilty, I am guilty, I am guilty.*

(I have mentioned Cortazar, Hawthorne, Poe; I haven't mentioned Valenzuela, Lispector, Maria Luisa Bombal, all of them ancestors who show me different paths. I have seen the approval of the man of the moment make the rounds many times, seeing me furiously splashing in an ocean of keys, distrusting. "What are you doing, baby? What's the rush to write? Do you want everyone to know what you ought to hide? You throw yourself out on to the street naked, undressed, like a crazy woman, like an easy whore. You shouldn't write about that, about everything that attacks you."

And if I don't write about this, then what? About public affairs? About society? About politics? I probably could never narrate those themes, for the distance between my body and the street becomes a greater abyss every day. Even the distance between my words and those of the other writers ... Now I wish this story was written in another way. That it would speak of colonies, that it would connect the history of the writer with her people, the women, the nation, with History with a big H, with the stories that appear in textbooks. Nevertheless, in this story there isn't space for more than the maps that mark my routes, those of the writer, and of Marina and Abnel. I try to understand why it doesn't come out any other way. I can't attain definitive answers. It would be a political impossibility, a lack of commitment that would let my voice follow the paths made by the bodies. It would be my fault ... Maybe that's why this story is so populated with questions and accusations.)

12:30 A.M.
The writer entered her room and closed the door. The kids were asleep. She was dead tired. Tomorrow she had to go to the office, take her history class, the last requirement for her master's, and then meet with her counselor to develop her study plan for the final comprehensive exam. And then arrive home, around six, wait for the ex to bring the kids from daycare, and try to imagine the story's ending. Exhausted, she went to wash her face, to gather her hair up in a ponytail. She went to wash away the spectres of the story that she knew had tracked her to the bedroom.

She had to sleep, really sleep, without nightmares, or attacks of illumination, or any other bullshit. She approached the bed. Sleep. Sleep. Hoping that the train of unconsciousness will run over her, that they'll perform a permanent frontal lobotomy on her, that the kids are grown up already and out of the house, that an envelope will arrive with the master's already granted, that dinner will cook itself, that Abnel will let Marina tear out the little baby so she doesn't have to go through what I have. And it's not that I don't love the children, they are people, how could I not love them? But I want to love them on weekends, on Tuesday afternoons, at the graduation, when they move away from home and they visit me and I told him, I told the ex that I didn't want custody, that it had been enough to give birth to them, to breast-feed them (the photos the bastard took of me, pulling out my nipples in public), calling me *malamadre*, and me mute like an idiot. What was happening to me all that time? What the hell did I have in my head? Sleep, sleep, raw ice, raw blood, a raw block of cement in my head, knocked up, at 21 PREG-NANT like Marina and when the grown-ups entered into it, they organized an automatic wedding for me, ultra-pompous, and I let myself be led to the slaughterhouse, I the rebellious one, the storyteller, the hairy one, but not Marina. Marina's going to tear out the little baby; she knows she's doing it for me, so I can sleep at night. Hmmm, calm down now, to sleep.

The writer dreamed even though she didn't want to; she dreamed she was seated in front of the other woman. The other woman was her (was me), in the study, the other, but without a face, and the absent face was like paper, a paper with writing on it. It read *Marina Marina Marina Marina Marina*. Someone walked in. He began to touch her breasts. She didn't see him, he was behind her, his shadow reflected on the breasts of the other one. He touched them, gentle, warm hands unbuttoning the dress, breasts out in the air. The other one tilts her neck back, the desire that arrives, they kiss her on the neck, she stretches out in the chair, almost without moving, in a slippery way. They touch her, they touch them; she doesn't see who it is, the other woman with a face made of paper *Marina, Marina, marinamarina-*

marinamarinamarina, the belly, below the hands that stop on the pubis playing with little hairs, with hanging locks, she slips, the chair falls backwards. She can't see. *Marina, Marina*—Where will you be? Now she's not in the garage, she returned home, she got into the ex's (Abnel's) car. They were in the back, she didn't see them, only the reflection of the mumbled sound, the complaints, they were going at it in the back seat of the car. It was her ex-husband with someone, she heard his voice, she didn't see him. She adjusted the rearview mirror and it reflected images, gynecological shots of a tube of flesh, from the dick to the lips, inside, outside, the vulva pink like a pink poppy flower like the ones in the garden but more brilliant, she got hot and suddenly the paper face, *marinamarinamarina* it was her, the other woman again in the garage, the hands had gone away. A thread of blood came out of the other's nose. They had hit her. What will I have done? The thread poured out of hr nose; she sat still staring, the thread running, falling down the chin, neck, chest, arm, hands, thighs, legs, foot, slipping and the other one silent, will they have killed her, what will I have done? the thread going down, sliding like a red serpent, menstruating a path, towards her, Ah, No! What is this? It was coming towards her and no matter how hard she tried to move from the chair, no, she was stuck there, nailed there, there and the thread, touching her toes, and now, No!, legs and knees. She was naked, the slow thread, thighs, she stuck out her tongue to lick the corners of her mouth, No!, slowly and across her bellybutton now, and across the stomach with stretch marks, slowly, but without stopping, No!, I struggle, it's arriving at my nipple, and it rises, up from my breasts, and it traces my neck, hot like a tongue, and it crosses her jaw entering, entering her ear, No!, her ear. The writer woke up panting, touching herself, shaking out her ear, wanting to tear it off.

3:00 A.M.

"Abnel thinks and thinks and Marina speaking to him alertly, now awake and finishing her Marlboro Light, now a little more clear-headed thanks to her explosion of words. That always worked for her, talk, talk, talk ... It awoke a hidden consciousness in her; she got more alert, more in touch with the intuition that told her things, that usually she edited. For four months she had been finding out secrets that Abnel thought he had been hiding well. Now she suspected, in this altered state of consciousness, the meticulous plan that the guy had been organizing for his escape, like a little game of chess. The hyper-preconceived caresses testified to it, the plans to convert themselves into the characters in Marina's stories, make them flesh and blood, make himself the lover of a thousand faces so she wouldn't have to invent them alone. She smelled it from the moment when Abnel invited her to mold him as if she were writing him.

"Marina hadn't wanted to hear him before. Ah, but this time things had turned serious. It was about killing a child. Because of that, she united all of her consciousnesses, and they made her move her legs silently to where Abnel was still clutching the bathroom sink. She started to talk, this time not to cleanse herself, but to camouflage her objectives. She was sure that by seeing Abnel's face just once without the make-up of a sad lover, she could corroborate those plans that her subconscious had suspected. So Marina went over there talktalktalktalk 'And if you knew how freaked out I was, but I think I'm resolving my doubts. The shock is diluting bit by bit; you don't know what good words do for me, talking to you does for me. Just as soon as I leave here, something tells me that I'm going to feel much better. I won't feel like I've felt the last few weeks, so confused, so changed, and suspecting that in reality something is happening to me and I don't know, I don't know what it is, but I'm finding out.' Walking silently towards Abnel who's shipwrecked, lost, scheming, trying to improvise as best he can. 'Now even the old folks say it, there's no problem that doesn't bring its blessing,' sliding like a murderous cat until she can uncover the eyes of the enemy lover, his mouth gathered into a pout, his sad and failed mustache, his distressed back and his crippled chest, his hanging butt, covered with pimples and scratches, his minuscule and curled-up penis, like it was asking forgiveness, as though it was trying to convince her that she wasn't the cause of so much confusion. Talktalktalk Marina and discovering Abnel Nieves Contreras, draftsman for a firm of agronomists, then recognizing herself as the one she always was, Marina Segarra Ortiz, literature student and aspiring storyteller. She had also played an unfortunate game. She had also represented herself superficially. And to recover her density, What a shame!, she would have to end up hating Abnel Contreras and poor Abnel wouldn't have any option but, along with his scion, to die."

Marina asks him for half of the money for the abortion and he gives her enough to pay for it completely. She takes it. Abnel makes a joke that she doesn't even hear. She dresses, now convinced that she has to get out of there as quickly as possible. She puts her clothes on with an incredible precision, thinks with an incredible precision, assumes control of her life and of what is happening to her with an incredible precision. That's it, first one leg of the pants, the other, and then the zipper.

(First this sentence, then the other, now the story is taking its place like Marina's pants. There are still many unresolved things, loose strands, suggestions, like Marina herself trying to understand how she managed to fall into this same ancient trap of converting herself into the superficial bitch and lover. Just like the writer in her hot studio trying to understand how it is that one story, an insignificant story, has helped her assume with precision a character giving her life, step by step to go out and have an abortion, tear a fictional creature out of herself that she doesn't need anymore, light and ready to throw herself down the routes of other stories. First this sentence, then the other, everything falls into place.)

Go to the bank to cash Abnel's check. Call Nandita so she'll accompany her tomorrow at eleven to extract the baby. Take a long slow shower to erase those saliva traces from her skin forever. Throw out all the stupid little things that Abnel had given her these last few months. Get the typewriter out of the closet. Buy paper. She knew that it wouldn't be now, that maybe years would pass, but this double abortion and re-birth surely would generate themes for a tremendous story.